1 MONTH OF
FREE
READING

at

www.ForgottenBooks.com

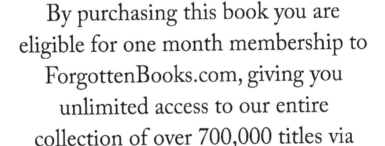

By purchasing this book you are eligible for one month membership to ForgottenBooks.com, giving you unlimited access to our entire collection of over 700,000 titles via our web site and mobile apps.

To claim your free month visit:
www.forgottenbooks.com/free178016

ISBN 978-0-267-48923-7
PIBN 10178016

This book is a reproduction of an important historical work. Forgotten Books uses state-of-the-art technology to digitally reconstruct the work, preserving the original format whilst repairing imperfections present in the aged copy. In rare cases, an imperfection in the original, such as a blemish or missing page, may be replicated in our edition. We do, however, repair the vast majority of imperfections successfully; any imperfections that remain are intentionally left to preserve the state of such historical works.

CONTRIBUTIONS

TO THE

PHYSICAL HISTORY

OF THE

BRITISH ISLES.

WITH

A DISSERTATION ON THE ORIGIN OF WESTERN EUROPE, AND OF THE ATLANTIC OCEAN.

ILLUSTRATED BY TWENTY-SEVEN COLOURED MAPS.

BY

EDWARD HULL, M.A., LL.D., F.R.S.,

DIRECTOR OF THE GEOLOGICAL SURVEY OF IRELAND, AND PROFESSOR OF GEOLOGY IN THE
ROYAL COLLEGE OF SCIENCE, DUBLIN.

LONDON:

EDWARD STANFORD, 55, CHARING CROSS, S.W.

1882.

INSCRIPTION.

TO MY VALUED CONTINENTAL FRIENDS

Dr. LAURENT-GUILLAUME DE KONINCK,

Professeur émérite à l'Université de Liége;

Dr. FERDINAND RÖMER,

Geb. Bergrath und Professor an der Universität, Breslau;

AND

Dr. ARNOLD VON LASAULX,

Professor der Mineralogie an der Universität, Bonn;

THESE PAGES ARE RESPECTFULLY AND LOVINGLY INSCRIBED.

E. H.

Dublin,
29th September, 1882.

PREFACE.

I HAD intended giving the title of "Palæo-Physio-graphy of the British Islands" to this volume; but certain friends in whose judgment I have the most implicit confidence assured me that no one would have the slightest conception from such a title of the contents of my book, and consequently that it would be useful only to adorn the shelves of my publishers, or of such of my private friends as I might feel disposed to present with a copy. I was aware that the term "Physiography" had been invented by an eminent naturalist, and I supposed that, having been popularised through the instrumentality of his pen, the reading public would have had no difficulty in discovering that "Palæo-Physiography" meant the physiography of past geological times. But I was assured that none but a few of the most profound Greek scholars would be able to elucidate the meaning of such a terribly long word; and as I was aware that, in the first place, one-half of the reading public never learn Greek at all, and that of the other half only about one per cent. retain anything of their knowledge after leaving the halls of *alma mater*, I waived my predi-lections in favour of the name at first proposed, and adopted the more commonplace designation which will be found on the title-page.

The volume itself, indeed, would never have seen the light had it not been for a fortunate concourse of circumstances which may be, in confidence, revealed to the inquiring public.

For some years past I had contemplated the preparation of a series of maps of the British Isles for the purpose of illustrating, as far as our knowledge and data would allow, the physical geography of successive geological periods; thus showing the changes of form through which our little isles have passed from their first genesis down to the period in which we have the happiness to live. I was deterred, however, by the view of the cost of publication. To carry out my intentions as I desired would require a series of chromo-lithographic plates, which would prove so heavy an expense at the outset that I feared no publisher would be sufficiently courageous to undertake the cost of publication. But what I despaired of having done through the ordinary channels I was enabled to accomplish with the aid of the Royal Dublin Society, in whose Transactions these little maps have already made their appearance. A fresh edition has been struck off for these pages, and it will give the reader some idea of the labour and cost of engraving when I state that no fewer than sixty-two lithographic stones have been used in their preparation.

To myself the labour of preparation of the plates has been one of love;—a gratification arising from the prospect of carrying out a project I had long entertained; and as I have received from various quarters some unexpected expressions of approval of this undertaking, I have reason to hope that those interested in problems of physical geography will find the book useful.

That the idea has occurred to others I am aware from the following interesting fact, which I may here be allowed to register. About a quarter of a century ago, the late Mr. William Longman, the then head of the eminent publishing firm of that name, proposed to my friend, the Rev. Dr. Haughton, Fellow and Professor of Trinity College, Dublin, that he should undertake the preparation of a set of maps of the British Isles, somewhat of the character of those now produced, with a view to publication. Circumstances prevented the execution of this work by one so well qualified to carry it out, but it may be affirmed with some degree of confidence that the time had then scarcely arrived in which the project could have been executed with that degree of exactness now attainable. Our knowledge of the *general* geological structure of the British Isles has not been very materially extended since that period as far as regards the surface distribution of the strata ; but another branch of inquiry, that regarding the interior structure, has received very important additions. This is due to the discoveries made during the last quarter of a century by deep borings in search of coal or other minerals, and of water, over extensive districts, but chiefly in the central and south-eastern counties. These borings have been of the greatest interest to geologists, because they have revealed the internal

structure of a large extent of country which would
otherwise have been the subject only of conjecture, or
of geological inference. For the present purpose
the details of these borings have been of material use.
Certainly without their aid it would have been impossible for me to have shown with the exactness now
attainable the underground limits of the Triassic,
Liassic, and Oolitic formations in the direction of the
Thames Valley and of the eastern coast, after they
have been successively lost sight of beneath the more
recent deposits.

Amongst those to whom I am indebted for light
thrown on the internal structure of the British Isles,
there is one name which takes unquestioned pre-
cedence—that of Mr. R. Godwin-Austen—in whose
elaborate essay, published in 1856 by the Geological
Society of London, many of the problems since
elucidated by experiment were investigated and
solved. In reperusing that essay by the light of subsequent discoveries, I have been profoundly impressed
by the knowledge of details which its author evinces
regarding the geological structure of the region of
which he treats, and the masterly manner in which,
by the handling of these details, he has enabled us to
understand the physical features of Western Europe
in past times. I feel persuaded that that essay will
ever be considered a masterpiece of geological
induction.

CONTENTS.

PART I.

CHAPTER I.

INTRODUCTION.

CHAPTER II.

THE GENERAL PRINCIPLES TO BE OBSERVED IN ATTEMPTING
RESTORATIONS OF OLD SEA-MARGINS.

CHAPTER III.

CONTINUITY: ORGANIC AND TERRESTRIAL.

CHAPTER IV.

CONTENTS.

CHAPTER V.

CONDITION OF THE ATLANTIC AND EUROPEAN AREAS IN GEOLOGICAL TIMES.

CHAPTER VI.

THE GENESIS OF THE NORTH ATLANTIC OCEAN.

PART II.

PALÆO-GEOGRAPHICAL AND GEOLOGICAL MAPS OF
THE BRITISH ISLES.

———

CHAPTER I.

THE LAURENTIAN PERIOD.

CHAPTER II.

THE CAMBRIAN PERIOD.

CHAPTER III.

THE LOWER SILURIAN PERIOD.

CHAPTER IV.

THE UPPER SILURIAN AND DEVONO-SILURIAN PERIODS.

CHAPTER X.

THE JURASSIC PERIOD (INCLUDING THE RHÆTIC, LIASSIC, AND OOLITIC DIVISIONS).

CHAPTER XI.

THE CRETACEOUS PERIOD.

CHAPTER XII.

THE TERTIARY PERIOD (INCLUDING THE EOCENE, OLIGOCENE, AND MIOCENE DIVISIONS).

CHAPTER XIII.

THE GLACIAL OR POST-PLIOCENE PERIOD.

ILLUSTRATIONS.

—◦◇◦—

WOODCUTS.

PLATES.

(MAPS OF THE BRITISH ISLES.)

CONTRIBUTIONS

TO THE

PHYSICAL HISTORY

OF THE

the several parts with the by-gone forms of the fauna and flora of successive periods. In this way, were

B

CONTRIBUTIONS

TO THE

PHYSICAL HISTORY

OF THE

BRITISH ISLES.

—◆—

PART I.

——

CHAPTER I.

INTRODUCTORY.

HOW WE ENDEAVOUR TO REPRODUCE TO OUR MINDS THE PHYSICAL GEOGRAPHY OF SUCCESSIVE GEOLOGICAL EPOCHS.

GEOLOGY is the geography of the past. As the geographer of the present day seeks both by the pen and pencil to produce a vivid picture of the physical features of some country, and to portray the characters and habits of the animals and plants which inhabit it, so is it the aim of the geologist to reproduce by similar agencies the past physical features of the earth's surface, to restore the outlines of land and sea, to recall the position of the mountains, valleys, and submerged depressions, and to re-people the several parts with the by-gone forms of the fauna and flora of successive periods. In this way, were

B

our means of information sufficient, we might present
for any special country, such as that of the British
Isles, a series of pictures, representing from very
early times the various phases of surface-sculpturing
which it has undergone, and the changes which have
succeeded each other in the animal and plant life,
whether aqueous or terrestrial, of the corresponding
periods.

It will be evident from the above statement that
the subject of Palæo-physiography (or the geography
of past geological times), divides itself naturally into
two branches, one dealing with the physical, the
other with the vital characters of special regions.
For a considerable portion of the earth's surface,
palæontologists have succeeded in reproducing the
groups of animals and plants which have peopled
the waters and decorated the lands of these regions
at successive stages of by-gone time. Perhaps of all
the attempts in this direction none have been so
successful as that of Dr. Oswald Heer, who has
graphically recalled for us the extinct fauna and
flora of Switzerland, tracing them down from forms
long passed away to those immediately preceding,
and closely allied to, those of our own times. With
no less skill has he also endeavoured to bring to our
minds the successive changes which the features of
that wonderfully diversified country have undergone,
corresponding to the changes observed in the animal
and plant life. American geologists have also done
much in the same direction, and with a firm hand
and appreciative eye, have portrayed the examples

of rock-sculpturing and illustrations of surface-change which nature has so profusely disclosed amongst the mountains, plains, and cañons of the Far West. Still, even in the restoration of the past life-history of any special region much remains to be done, from the fact that our knowledge of the animals and plants of geological periods is largely restricted to those forms which inhabited the lakes and seas, and whose remains are entombed in the strata that were deposited beneath their waters. Of the land plants and animals our knowledge is comparatively limited, and depends on such fragmentary remains as may happen to have been drifted out to sea by streams or currents, subsequently preserved, and ultimately brought to the notice of the observer.* The land forms of animal life during past geological periods, are probably destined ever to remain to a great extent unknown to us, and we must be fain to content ourselves with the rich, and often profuse, assemblages of the shells and skeletons of aqueous forms which are preserved for our study in the strata of the earth's crust.

Being thus driven back by the force of circumstances from attempts to restore in imagination the fauna of the land of geological times, may we not essay to accomplish something in the way of physical restoration? Here surely we can find some little solace in our disappointment; and, foregoing a line of

---

* The small proportion of land forms dredged up in the seas around the coasts of the British Isles, shows how entirely unrepresentative of such forms may be the fossil contents of the strata.

study from which nature has remorselessly debarred us, take up another for which she has furnished more ample materials.* If we are unable to people the lands of the Carboniferous, Triassic, Jurassic, and more recent periods with their successive forms of animal life, may we not, at least, attempt to ascertain where the lands of those periods lay over and around such a region (for instance) as that of the British Isles; where were the respective margins of land and sea, and where the areas deeply submerged, or, on the other hand, greatly elevated? Or, going into a still wider field, may we not seek to ascertain from the method of such an inquiry, whether any light is thrown on some of those interesting physical problems which are so intimately connected with biological study, such as the "permanency of oceans and continents," and the position of the lands from which were derived our sedimentary rocks? Nay, may we not even attempt to represent these details on maps, showing as in those of the present day of the same region, but with

---

* In reference to the question of the imperfection of the palæontological record, especially as regards land animals, Professor Haeckel has pointed to a source of such imperfection not usually recognised. He shows that permanent strata which enclose organic bodies can only be formed during the slow sinking of the ground under water; but when the submerged floor is being elevated, the strata which may be deposited are liable to be constantly swept away in shallow water. Hence we have no traces of the organic forms living during periods of elevation, although it can be theoretically proved that the variety of animal and plant life must have increased during these very periods of elevation. For, as new tracts of land are raised above water, new islands are formed. Every new island, however, becomes a centre from which fresh varieties have their origin; but owing to the causes assigned, their records are lost to us.—'The History of Creation' (Schöpfungsgeschichte), E. Ray Lankester's Trans., vol. ii. p. 23.

less detail and often doubtfully, the position of land and water, the heights of the one and the depths of the other, and thus trace through the long past of geological time the various phases of transmutation through which our little section of the globe has passed?

Such is the task I have set before myself in the following chapters; but before entering on a description of the maps themselves, and the details represented on them, it is necessary, in the first place, that I should state the general principles upon which these physical restorations have been attempted; and, in the second, the general conclusions regarding some of the problems above alluded to, which seem to be fairly deducible from these restorations.

CHAPTER II.

GENERAL PRINCIPLES TO BE OBSERVED IN ATTEMPTING RESTORATIONS OF OLD SEA-MARGINS.

(*a.*) *Formation of sedimentary strata.*—From the remotest period which can be strictly comprehended within the range of geological time, the crust of our earth has been undergoing change. Ever since the consolidation of the crust, and the condensation of aqueous vapour on the surface, a period of which we have no record amongst the existing rock-masses of the globe, land and sea have been changing places somewhere or other; and while this has been going on, the land surfaces in all districts, where rain has fallen and rivers have flowed, have been subjected to a process of waste, known amongst geologists by the term "denudation." The continuous action of rain, rivers, waves acting on coasts, snow and ice, and even winds, has been to denude (or lay bare) more and more of the land surface, and to transport the derivative material, in the form of gravel, sand, or mud, into the adjoining seas. The distance to which these materials are carried seawards varies greatly. As a general rule, the coarser materials are deposited near the shore, but the finer may be carried by the agency of ocean currents for hundreds or thousands of miles out to sea, before finally settling over the bottom. Nevertheless sediment thus derived is not usually carried into mid-ocean, a fact determined by deep-sea soundings,* but is deposited within a zone of a few hundred miles from the shore.† We may suppose, however, that in all cases the sediment becomes finer as it increases in distance from its source, and that it tends to thin away and disappear in the direction of mid-ocean.

The converse of this is (within certain limits) equally true, namely, that the sediment will tend to increase in thickness in the direction of its source; so that we have here a most valuable guide for the determination of the directions towards which we are to look for the lands which were the sources of strata during any special geological period. If, in the case of any formation composed of strata formed under water—such as sandstone, shale, and their varieties— we find that it gradually becomes attenuated towards the south, or, conversely, that it swells out in vertical dimensions in a northerly direction, we have *some* evidence at least (if not always conclusive) that the land of the period from which the sediment was primarily derived lay to the north of its position. The above is a principle of very wide application,

---

* First shown by Dr. Wallich from the soundings made in the Atlantic preparatory to laying the Atlantic telegraph cable; more recently by those of the *Challenger* expedition under Captain Nares.

† In some cases the distance is much greater, as, for instance, in that of the Amazon, the muddy water from which discolours the ocean at the surface for several hundred miles, and by the time the mud subsides, it must reach to much greater distances.

and we shall find when dealing with special formations over the area of the British Isles that it has
numerous illustrations.   It has been ascertained, for
instance, that some formations, such as the Upper and
Middle Carboniferous, persistently become attenuated
in certain definite directions, and tend to swell out in
those opposite thereto; thus pointing in a manner
not to be mistaken to the position of the old Carboniferous lands.

(*b.*) *Changes in the characters of strata.*—In addition
to strata of strictly sedimentary origin, such as sandstones, slates, shales, clays, and their varieties, derived
by mechanical processes from the waste of land,
geological formations are often composed entirely or
in part of limestones, the origin of which is widely
different from that of the former class of rocks.   In
general a close examination of specimens, especially
if aided by the microscope, will suffice to convince
the observer that both the material of these rocks
and their structure is essentially different from those
previously described.   They are found to be composed of carbonate of lime, sometimes in combination
with carbonate of magnesia, and other ingredients
not derivative in the form of sediment from the
waste of lands.   Further examination will in most
cases tend to convince him that they are essentially
organic in structure.*   They show structures corresponding in form to those of shells and skeletons of

---

* Except in the case of metamorphic limestones, which have undergone
a process of transformation, owing to which the original organic structure
has been changed into one more or less crystalline.   The marbles of
Carrara, Pentelicus, &c., are illustrations of the kind.

molluscs, corals, crinoids, star-fishes, echini, foraminifera, &c., such as inhabit the waters of the sea or
ocean, and often form the material of its bed over
the central portions, far out of reach of muddy or
sandy sediment. The examination of thin trans-
lucent sections of the limestones under the microscope
will often disclose organic structure where none is
apparent to the eye even when aided by the lens; and
the most dense and compact portions will be found
to display the forms of foraminifera, together with
fragments of shells, corals, or sponges; the cementing
material being impalpable calcareous mud derived
from the decay, or attrition, of similar bodies.*

Viewed as a whole, it becomes evident that such
limestone formations must have been built up in a
manner altogether different from that under which
the sedimentary strata were accumulated. They are
essentially organic in structure, and must therefore
have been constructed through organic agency.
They are more or less free from sediment—sometimes
absolutely so—from which we infer that they have
been deposited far from land, and in pure transparent
waters. They resemble the calcareous ooze of the
mid-Atlantic ocean, or the material in course of
formation inside the walls of an atoll, or of a barrier
reef; allowance being made for the differences in the
forms of the animal structures belonging to past
geological times as compared with those now living
in the sea or ocean. Wherever, therefore, we find

---

* Spicules of sponges, shields of Polycistineæ, and other siliceous forms
are frequently associated with those of calcareous composition.

great beds of limestone of hundreds or thousands of feet in thickness, not associated with beds of sand-stone, shale, or clay, we may feel sure that they were formed far out from the lands of the period to which they belong, and are more or less deep-sea deposits.

(c.) *Interchange of deposits of organic and mechanical origin.*—If we confine our attention for a moment to any considerable oceanic region, such as that of the North Atlantic, it will be evident on consideration that very dissimilar deposits are being contempora-neously laid down over different parts of its bed. Thus in the central portions calcareous mud formed of the shells and skeletons of marine animals, chiefly foraminiferal, is in course of deposition; while round the coasts and opposite the mouths of rivers, muddy and sandy sediment passing landwards into shingle is being laid down. But between these two regions, the central calcareous and the marginal sedimentary, there lies a "debatable land," so to speak, of varied materials constantly mingling or inosculating, and probably of considerable breadth in some places. Considered in relation to vital action, a perpetual struggle between the organic and inorganic agencies is in progress over this area, arising from the in-jurious effects of sandy or muddy sediment on the vitality of many of the forms which may be called "limestone builders." It is known from the obser-vations of Ehrenberg, Darwin, and Jukes, that most of the existing polyps are destroyed by the presence of sediment in the sea, and that it is only in clear uncontaminated waters that they are capable of

carrying on their vital functions. The same observation applies with more or less force to other forms
of marine life, such as crinoids and foraminifera, and
in a still less degree to some classes of the mollusca.*
Hence we must suppose that in the border zone
between the regions of calcareous and sedimentary
deposits, there is a constant advance and retreat on
the part of the limestone builders according as the
waters of the sea become more or less pure. Now,
such changes in the character of the waters over the
tract referred to, must from time to time recur in
consequence of changes in the physical geography of
the adjoining land and sea areas. In some places
the land is gaining on the sea, and the areas of
drainage are being proportionately extended; in
such cases the amount of sediment carried down into
the sea will tend to augment, and will influence the
character of the waters, which in turn will have a
reflex action on the deposits going on over the sea
bed. Other causes, such as the temporary or permanent deflection of tides and currents, will bring
about similar results, and according as the sea water
so affected becomes favourable or otherwise, organic
action will be quickened or retarded.

Upon such principles we may explain the occurrence of alternate beds of limestone or shale and
sandstone amongst some formations, or the entire

---

* It is true that foraminifera and crinoids are found alive near our
coasts, but the extent to which they contributed to the structure of the
great and pure limestone masses—such as those of the Carboniferous,
Cretaceous, and Tertiary periods—show that they flourished only where
the waters of the sea were free from sediment.

replacement of strata of one kind by those of another; as was the case when the Carboniferous limestone gave place to the shales and sandstones of the Middle Carboniferous group. Our special purpose here is to note the bearing of these changes on the question of the relative positions of land and sea. It will be clear from what has been said—that in the direction where a limestone formation becomes more or less intermingled with beds of sandstone or shale, or is entirely replaced by them, in *that* direction may we look for the position of the land whence these materials were derived; while, on the other hand, where the formation is free from such ingredients we may assume the area of deposition to have been remote from land, or at least free from the presence of mechanical impurities. On such grounds we have reason for believing that during the epoch of the Carboniferous limestone considerable tracts of land existed to the north and west of the British Isles.*

(*d.*) *Shelving sea-bottoms and coasts.*—On the other hand, it must not be forgotten that all formations tend rapidly to thin away in the direction of any tract of contemporaneous land, and its bordering shelving sea-bottom, though not of sufficient extent to be in itself a source of much, if any, sedimentary materials. A rock, or island, rising out of deep water, and around which calcareous or sedimentary strata are being deposited, will necessitate the wedging out of such strata, as the water becomes shallower on approaching its margin; and if the

* Plate VI.

rock or island be of small area, very little alteration will take place in the nature of the deposit up to within a short distance of the shore. Of this we have a good illustration in the case of the old ridges of Lower Palæozoic rocks which stretched across the centre of England during Carboniferous and Permian times, and separated the strata of these periods into two areas of deposition ;* and in the concealed ridge of Palæozoic rocks which occupied the region of the Thames valley and the south-eastern counties of England during the Jurassic period.† Against the shelving shores of these ridges the newer formations successively wedge out, with but very little change in their composition.‡

(e.) *Shore beds.*—The presence of beds of shingle or conglomerate generally indicate marginal conditions. These are the materials laid down at the mouths of rivers, or broken off, rolled about, and rounded by wave action along the coast of the adjoining land, and strewn over the sea bed to greater or less distances from the shore. Such shore beds are seldom found unless in cases where strata are being formed in the vicinity of rocks which have been tilted and upraised somewhat abruptly in relation to the adjoining sea bed. The British formations afford numerous illustrations of such littoral deposits.

---

* Plates VI. and VII.　　　　† Plate X.

‡ Thus the Carboniferous limestone of Breedon Hill and Ticknall, near Ashby de la Zouch, preserves its usual character, though only a short distance from the margin of the very old rocks of Charnwood Forest, which must have formed a ridge of unsubmerged land, or at least a series of rocky islets, a short distance to the southwards.

The Cambrian conglomerates of the north highlands of Scotland are deposits of this kind on a large scale. On a scarcely less massive scale are those of the upper Llandovery stage of Ireland and Wales, of the " Lower Old Red Sandstone " of central Scotland and the north of Ireland, and the Old Red Conglomerate of the south of Ireland. Shore beds very characteristic, but less remarkable in extent, are found amongst the Triassic and Permian strata of the midland and western counties of England and the Cretaceous beds of Wilts and Berkshire.* It is remarkable, on the other hand, to how great a distance rounded pebbles may be carried from their original sources; as, for example, in the case of the quartzite pebbles of the New Red Sandstone of the midland counties of England, the source for which must apparently be sought for amongst the Old Red Conglomerates of the south of Scotland;† and those of Budleigh-Salterton, in Devonshire, derived from the Palæozoic beds of Normandy and Brittany.‡

(*f.*) *Discordancies of stratification.*—In determining the position of the land areas and adjoining seas during successive geological periods, one of the most important guides to be observed is the discordant, or unconformable, relations of successive groups of strata.

---

* The conglomerates formed at varying geological horizons amongst the Devonian beds of Belgium, near their junction with the older rocks of the Ardennes, are excellent examples of shore beds, but are somewhat outside the area of our inquiry.

† ".The Triassic and Permian Rocks of the Midland Counties," Mem. Geol. Survey (1869), p. 59. See note, p. 98.

‡ Vicary, Quart. Journ. Geol. Soc., vol. xx. p. 283 ; Edgell, *ibid.*, vol. xxx. p. 45.

Where, too, such groups, as for example those of the
Triassic and Silurian periods, are in contact, there is
much probability that the older formation has in some
region occupied the position of a land surface, while
the newer was being deposited under the waters of
the adjoining sea, or inland lake. In such a case the
actual margin is frequently indicated by the abrupt
uprising of the older formation with reference to the
newer, as shown in the case of the Trias of the Vale of
Clwyd, in North Wales ; but it must be recollected that
during deposition the margin was constantly changing
its position, and as time went on crept gradually over
the flanks of the subsiding land during the slow depression of the area of deposition. If the two unconformable sets of strata are in immediate or close geological sequence, then the marginal relations are more
easy of determination than when it is otherwise. Such
is the case, for instance, with the Lower and Upper
Silurian strata, which, though highly discordant as
regards stratification, are in immediate geological
sequence in these islands. The original shore beds,
those of the "Llandovery stage," are found in Shropshire, Wales, and the west of Ireland in the form of
conglomerates or pebbly beds not far removed from
their original position of shore beds.* The position
of such conglomerates, however, when in actual

---

* Elsewhere I have shown that the hiatus between the Lower and Upper
Silurian beds was sufficient to include the time necessary for, not only the
disturbance and denudation of the beds of the older formation, but for
their metamorphism, as the fossiliferous Llandovery beds are found in
Connemara to rest on an eroded floor of metamorphosed Lower Silurian
strata ;—a fact recognised by the late Prof. Harkness.

contact with the strata of the older formation, indicates their relation to each other at some special stage of deposition. As the older rocks were more and more depressed beneath the waters, and the submerged area was being filled in with fresh materials, the margin would be extended and rise, so to speak, higher and higher on the subsiding land surface. It is owing to subsequent denudation that the actual position of the line of contact has been revealed to us. At a subsequent geological period, the whole of these strata were upraised, tilted, and exposed to rain and river action, by which, in process of time, hills of considerable elevation, such as Mweelrea, have been sculptured, and deep channels have been carved out, such as the remarkable fiord known as " Killary Harbour," together with numerous other hollows and channels, amongst which sections of the strata illustrating the relations of the older to the newer are laid open to view.*

* 'Physical Geology and Geography of Ireland,' pp. 21–3.

CHAPTER III.

CONTINUITY : ORGANIC AND TERRESTRIAL.

(a.) *Of the grouping of strata and their organisms.*—
The principle of continuity pervades terrestrial, as
well as organic, evolution ; life has never ceased on
the globe since it was implanted therein by The
Creator, neither has deposition of strata been completely interrupted for a moment. But in order to
illustrate the succession of animal and plant life
naturalists have seized upon those forms which are
characteristic of successive geological epochs. To
present in unbroken succession the forms of animals
and plants from the earliest dawn of life to the
present day, would be impossible from the materials
within our reach. The fauna and flora of successive
geological periods differ, more or less, from those
which precede and follow them ; but the intervening
forms by which they were connected, and through
which they passed, have been lost to us. For
purposes of illustration and description this natural
arrangement is exceedingly convenient. Were we
acquainted with the forms and characters of the
whole unbroken chain of life in any one province,
we should be liable in our attempts at description of

C

orders, genera, and species, to fall into chronological error, and to place forms side by side which were really not synchronous. From this we are preserved, owing to the natural grouping of strata into systems and formations, which over special regions are either synchronous or at least homotaxial; and which with their organic contents are divided off from each other by local breaks in the continuity of formation, resulting in unconformities of stratification representing in each case a more or less prolonged lapse of unrepresented time.

This natural grouping of animals and plants by formations, or successive epochs in geological history, depends on those physical disturbances which, from time to time, have broken in upon the slow and gradual accumulation of strata over the bed of the sea, and which have resulted in the extinction of genera and species over special provinces, and their subsequent replacement by immigrations of new forms from other regions of the ocean upon the return of tranquil conditions of deposition. It will thus be seen that the periodic appearance over limited regions of successive assemblages of plants and animals is due to the recurrence of purely physical causes acting over such regions, and is dependent on them. The breaks in the continuity of organic forms are due to the breaks in the continuity of stratification, owing to which it becomes possible to represent the geographical conditions of any special region at distinct epochs in a manner somewhat parallel with those characterised by distinct assemblages of organic forms.

It is, indeed, entirely owing to such local breaks

in continuity, that geologists have been enabled to arrange all the strata of the globe into divisions (such as " systems," " formations," " groups," &c.) sufficiently definite for classification, each characterised by special forms of plants and animals; and in endeavouring to prepare a series of maps representing the palæo-geographical features of some region, such as that of the British Isles, the requisite number of such maps and their proper order of succession is already in a manner pre-determined, as it necessarily corresponds to those of the successive geological formations, or natural groups of strata.*

Thus, in preparing the following series of maps, illustrative of the past physical geography of the British Isles and of the adjoining portion of the European continent, I have not been left to arbitrary choice out of an infinite succession of phases through which the outlines of land and sea have passed between the Laurentian and the Tertiary periods. I have been necessarily more or less restricted, both as to number and stages, and have endeavoured to show the relative position of land and sea as they appeared at certain well-recognised epochs, represented by special geological formations which are laid down in the duplicate maps in each case by special colours.† These duplicate maps show the

---

* It might have been possible to prepare a larger number of palæo-geographical maps representing intervening stages than those accompanying this volume, and on a much larger scale it might have been desirable. But with the scale here adopted it will probably be admitted that the number is sufficient for purposes of illustration.

† Generally the colours used by the Government Geological Survey of the United Kingdom.

area occupied by each formation both at the surface and where overspread by newer formations, a lighter shade of the same colour being used for this latter purpose. One of the first and most obvious uses of these duplicate maps is to show the present, as compared with the original, superficial extension of each formation. It may be assumed that the original area of the formation in each case was approximately that overspread by the waters, whether of the sea or inland lake, under which the strata were deposited. These hydrographical areas are in most cases vastly greater in extent than the formation of the period as at present represented. The latter is, indeed, in many cases merely a fragment, or group of fragments, of a once wide-spread expanse of strata; and the question arises : how has this breaking up and destruction of such extensive tracts of originally horizontal courses of sandstone, limestone, shale, &c., been brought about ?

(b.) *Periodic disturbances of strata.*—The answer to this question must be more fully sought for in treatises on geology—of which, thanks to the popularity of the science, there is no stint—and cannot be more than glanced at here. Briefly stated, we may say that all aqueous rocks have been originally formed in approximately horizontal layers over gradually subsiding areas of deposition, but that from time to time terrestrial disturbances, the result chiefly of the contraction of the earth's crust due to secular cooling, have caused the strata to be upheaved into dry land, to become tilted at various angles from the horizon, folded, flexured, or even inverted. Such changes have been necessarily accompanied by fracturing and " faulting," and the strata once raised into land surfaces have been subjected to the action of sub-aërial denuding agents, such as sea-waves, rain, rivers, snow, ice, and winds, by which they have been worn down into valleys, river channels, and cañons; sculptured into ridges, cliffs, and escarpments; or raised into mountain peaks and precipices, together with those ever-varying forms of surface feature and contour by which our world has been so richly diversified. The materials thus drained off from the land have been carried by the rivers into the sea to form new strata, and have sometimes been used over and over again.

(c.) *Volcanic rocks.*—Vulcanicity has also played its part in the formation of rock-structures. Vast masses of molten lava have been poured out over the bed of the sea, or the surface of the land, either from vents of eruption, or from fissures riven through the crust.* Along with these molten masses are often to be found beds of volcanic ashes, lapilli, and agglomerate, which have been blown out of the craters of eruption and have been strewn over the surface of the land or the bed of the ocean.† These volcanic products, when elevated and subjected to atmospheric erosion, generally assume bold, craggy, terraced, or precipitous features.

* As shown by Prof. Geikie to be the case in the volcanic region of the Yellowstone and Madison rivers of Western America.—*Nature*, Nov. 1880.

† Some of these marine beds of volcanic ash enclose fossil shells, as in the case of those forming the summit of Snowdon.

(d.) *Metamorphic rocks.*—Of still greater importance is the group of crystalline, though stratified and foliated, rocks known as " metamorphic." Originally in the form of ordinary aqueous strata, they have been converted, by the combined agency of heat, pressure, and superheated water or steam, into crystalline masses similar to granite, or various forms of schist, quartzite, and marble. The process of metamorphism has been carried on underneath great masses of once superimposed strata, so that the presence of such rocks at the surface is due to subsequent upheaval and denudation. The metamorphic rocks are generally found amongst the mountainous districts of the world, and in the British area are the chief constituents of the Scottish highlands, together with those of Donegal, Mayo, Galway, and Wicklow in Ireland.

Of such materials then are the mountains, hills, and plains of the British Islands constructed. Acted on by terrestrial forces from below, and erosive agents from above, they have been moulded and sculptured into the endless varieties of hill and dale, mountain, crag, and fell, undulating plain, or gently sloping vale, which enable us to find, where'er we wander, something not seen before; some new feature in the landscape—something that reiterates in our mental consciousness—" Nature never repeats herself."

CHAPTER IV.

OF THE " PERMANENCY OF CONTINENTS AND OCEANS."

THE doctrine of the " permanency of continents and oceans " is one which has recently risen into favour, and has been advocated by several distinguished naturalists. Their view seems to be that the existing oceans occupy hollows and the existing continents ridges that became so some time after the consolidation of the earth's crust, and that they have in the main retained their relative positions down to the present day; or at least, that their relations have never been actually reversed. In the words of a recent writer : " The present continental ridges have probably always existed in some form, and as a corollary we may infer that the present deep ocean basins likewise date from the remotest geological antiquity." * In support of this view it has been stated on the authority of the late Sir Wyville Thomson that " few of the rocks known to geologists correspond exactly to the deposits now forming at the bottom of our great oceans ; " that the formations with which we are acquainted were for the most

* Geikie, ' Geological Sketches,' p. 324. The same view is maintained by Mr. Wallace, ' Island Life,' &c.

part deposited under shallow waters, and comparatively near to the shores of ancient lands.

In reference to the first statement it appears strange that upon the discovery of the " calcareous ooze " of the mid-Atlantic ocean, and the observation of its essentially foraminiferal composition,* its resemblance to the Chalk formation of the Cretaceous age was so generally recognised that some geologists went the length of affirming that we were still living in the Cretaceous period! From this pole of wild speculation we seem to be veering round to one of an opposite direction in the statement that the Chalk is merely the detritus of coral reefs formed in water of no great depth. We may, however, confidently affirm, that whatever be the nature of the marginal representatives of the Chalk—such as the Faxöe beds of Denmark—coralline structure is rare in the central portions of the Chalk area, extending from the north of Ireland,† through the south of England, into Belgium and France ; how much farther I cannot say. Whether studied microscopically, or on a large scale, its leading characters seem to be essentially oceanic, and prevail from Ireland on the west to the shores of the Caspian on the east, and from Belgium into North

---

* The first samples were brought up during the deep-sea dredgings made across " the Atlantic plateau " for testing the character of the bottom previous to laying the first Atlantic cable.   The organisms were described by Dr. Wallich, naturalist to the expedition, in the pages of the ' Quarterly Journal of Science,' vol. i., and elsewhere; Ann. Nat. Hist., vi. p. 457. See ' Voyage of the *Challenger*,' vol. i. p. 206 ; ii. p. 291.

† The foraminiferal structure of the Chalk of Ireland can be well studied under the microscope by means of thin sections, the rock being generally sufficiently hard for slicing.

Africa. These may be briefly defined as follows: Extreme purity, and freedom from admixture with foreign ingredients; regularity of bedding; the alternation of layers of siliceous bands and nodules, generally replacing sponge structures ;* the vast extent of its original area of deposition, which, continuing eastwards from the borders of the Atlantic, embraced all but isolated portions of the British Isles, the region of Belgium and of France, except the central granitic plateau; nearly the whole of the Iberian peninsula and of central and southern Europe, the Mediterranean basin, and vast tracts of North Africa, of Asia Minor, Palestine and the region bordering the Euphrates; a deposit which has been elevated to thousands of feet on the flanks of the Alps, the Pyrenees, the Caucasus, and the Himalayas, indicating vast changes of level since its formation. All this points to the existence during the Upper Cretaceous period of wide oceanic conditions, far removed from large rivers bearing down sandy or muddy sediment; and the consequent remoteness of continental land from the original Cretaceous basin of Europe and western Asia.†

If then it can be shown that during the Cretaceous period the centre of Europe as far north as the borders of Scandinavia, and the whole of the region fringing the Mediterranean on both sides and extend-

* As the late Dr. Bowerbank has demonstrated.

† The nature of the Cretaceous deposits themselves reminds us of those alternating calcareous and siliceous deposits, due to organic agency, now in course of formation over the floors of the Atlantic, Pacific, and Indian Oceans, as determined by the soundings of the *Challenger* expedition.

ing eastward for an indefinite distance into the continent of Asia, was submerged, except where a few islands and reefs of older rocks rose above the general flood of waters, there must have been (according to all the laws of terrestrial mechanics) a corresponding interchange of terrestrial conditions. In other words, a large area of the North Atlantic stretching from Scandinavia westward to the north of the British Isles into mid-ocean, may be presumed to have existed during the same period as dry land; or, at least, as a region only partially submerged.

A similar argument may be applied to the case of the Nummulite limestone formation of the Eocene period.*   Its characters are in the main analogous to those of the Chalk, and the vast area of land surface which it now covers, or underlies, attests the wide extent of the ocean waters under which it was originally formed.   But if so much that was ocean is now land, we may infer on the principle of interchange of physical conditions, that during the early Tertiary epoch, much that was then land is now the ocean.   In more ancient geologic periods, as we shall have occasion to see, the evidence points with still greater force in a similar direction.

* The Rev. Dr. Haughton gives the following elevations for the Nummulite limestone :—

Dent de Midi (Pyrenees)	10,531 feet.
Diablerets (Alps)	10,670 „
Western Thibet	16,500 „

and adds : " The Nummulite limestone formation extends through 98° of longitude, and is found in latitudes ranging from 15° N. to 55° N.," and he argues that " the great east and west chain of Europasia is the most modern of mountain chains " (' Lectures on Physical Geography,' p. 51).

CHAPTER V.

CONDITION OF THE ATLANTIC AND EUROPEAN AREAS
IN GEOLOGICAL TIMES.

LET us now inquire what is the evidence on this interesting question to be derived from the construction of palæo-physiographical maps of the British Islands? It must be admitted in the outset, that the British Isles occupy but a small area as compared with that of the Atlantic Ocean on the one hand, and of Europe on the other. Still, these islands are intimately connected as regards their geological structure with western and northern Europe, while they lie opposite some of the deepest abysses of the Atlantic. They are, therefore, conveniently situated for the purpose of observation and comparison, and any information they afford is valuable as a contribution to the general sum of our knowledge.

I propose, therefore, to pass in review some of the leading formations, beginning with the most ancient, and to discuss the question how far their nature, extent, and distribution tend to throw light on the question—what amount of antiquity are we to assign to the Atlantic Ocean?—reserving for future pages the details which must be here more or less anticipated. We commence, then, with the oldest known British rocks, generally known as " The Laurentian."*

(a.) The Laurentian beds crop out at various points along the north and west coasts of the British Islands, while they form a very large portion of the Scandinavian promontory. They reappear in central Europe, possibly in central France,† as well as in the borders of the Red Sea ‡ and the Upper Nile valley. The formation may therefore in one form or another be supposed to underlie almost the whole of Europe and the British Isles;—the truly " fundamental " § granitic and gneissic basis of all other formations. The thickness of this great group of strata must be enormous. We have been unable to discover either its basement or its culminating beds. Like time itself, we see, when traversing the Laurentian region of Scotland, for example, no trace of a beginning, no prospect of an end. At the lowest estimate the formation is over 20,000 feet in thickness, and may originally have reached over 30,000 feet.‖

Now, these beds, consisting of gneiss, hornblendic and micaceous schists, and other crystalline strata, are undoubtedly derivative. They once existed as sandstones, shales, clays, and limestones, which have since been subjected to intense metamorphism. They

---

* Plate I. Other names are " Archæan " and " Pre-Cambrian."

† It seems extremely probable that the granitic gneiss, and schists of central France and of Brittany, are referable to Laurentian age.

‡ As Lartet has shown, the Egyptian granite is in all probability referable to this formation.

§ The name applied by Murchison to this group of rocks in Scotland.

‖ In Canada the thickness has been estimated by Sir W. Logan at 35,000 to 40,000 feet (Geol. of Canada, p. 45).

were accumulated under the waters of the Archæan ocean from the waste of lands, which we conclude must have been of continental extent in order to supply materials so wide in their distribution, and so vast in their vertical dimensions.

It is also to be recollected that the Laurentian beds are quite as largely represented on the North American continent as they are in Europe — the larger part of Canada being formed of them*—while they pass below the more recent formations of the United States south of the St. Lawrence. Here, then, we have two vast tracts lying on either side of the North Atlantic formed out of a presumably contemporaneous group of strata of vast extent and thickness. Where, if not in the Atlantic itself, are we to seek for the place of the continent which supplied the quantity of material required for the construction of this venerable pile? To suppose that the strata may have really been shallow-water deposits, accumulated not far from the land of the period, offers to my mind no solution of the difficulty. If the Laurentian basin were depressed 35,000 or 40,000 feet during the deposition of the strata, by so much at least must the region of the Atlantic have been raised in order to remain as a land surface, and as a source of supply for the Laurentian strata. The actual rise may have been much more than this; but a rise of so much would go far to convert the ocean into a continent if it had been as deep as at the present time.† The conditions

* Logan, Geol. of Canada, p. 42.

† The soundings made by the officers of the *Challenger* in the mid-Atlantic descend to 3875 fathoms (23,250 feet), those of the U.S. Survey,

necessary for Britain and northern Europe being also necessary for North America, the same continent, extending probably into the Arctic circle and embracing Greenland, would necessarily be the parent land of the Laurentian strata on either side. This was the primæval Atlantis, rising probably into lofty mountains stretching far south of the 50th parallel, and bounded on either hand by the waters of an ocean which rolled over the regions of the North American continent on the one hand, and of the European on the other. At this period of the earth's history the North Atlantic Ocean had no existence; such, at least, is the conclusion to which I am impelled from the foregoing considerations.

(*b*.) *The Cambrian period.*—With the Cambrian epoch probably appear the earliest vestiges of the British Isles. After the formation of the Laurentian strata, they were metamorphosed, intensely flexured and contorted, depressed into furrows, which were occupied by the waters of the sea or of inland lakes, or raised into corresponding ridges.

A rib of these Archæan rocks appears to have been protruded above the waters of the ocean along a tract stretching in a north-easterly direction through the west of Ireland and the highlands of Scotland.* A parallel ridge probably uprose along the line of the outer Hebrides, and between these a basin seems to have been enclosed in which sandstones and con-

---

given in Maury's Phys. Geog. of the Sea, descend to 4000 fathoms (24,000 feet), this is probably a maximum. The average depth of the North Atlantic Ocean is 15,000 feet, nearly the elevation of Mont Blanc.

* Plate II. See also Quart. Journ. Geol. Soc., May 1882, p. 210 *et seq.*

glomerates were accumulated during the Cambrian epoch. These ribs may be regarded as the primary framework upon which the British Isles were constructed. One of them, that of the central highlands, may be called "the primæval backbone of Britain and Ireland." Their general trend was N.N.E. and S.S.W., and they were the outcome of lateral pressure due to the contraction of the crust in N.W. and S.E. directions over this region of the globe. It is probable the Atlantic continent was throughout its extent converted into a succession of similar ridges and furrows. To the east of the Archæan ridge, or backbone, above described, the ocean stretched away for an indefinite distance, overspreading the region now forming the eastern parts of Ireland, the whole of England and Wales, and large tracts of western and central Europe.* Thus far, at least, it cannot be said that the Atlantic as an ocean or Europe as a continent had come into existence.

(c.) *Lower Silurian epoch.*—Whatever incipient traces there may have been of the framework of the British Isles during the preceding epoch, they seem to have disappeared altogether beneath the waters of the Lower Silurian ocean which overspread the region of the British Isles, and stretched away over that of Europe, including much of Scandinavia.†

* It is now known that the rocks of the Ardennes, as well as some of those of Brittany, are of Cambrian age.—G. Dewalque, 'Description Géologique de la Belgique,' p. 18; Mourlon, 'Géol. de la Belgique,' t. i. p. 31.

† The Silurian strata occur in the south of Sweden and the environs of Christiania in a position nearly horizontal and discordant to the older rocks. Durocher, 'Annales des Mines,' t. xv., 4me ser.

It was an epoch of prevalent marine conditions where now dry land holds its own, and there must therefore have been a corresponding prevalence of land conditions where now the waters of the Atlantic profoundly roll.   The strata of Lower Silurian age are of marine origin and of great thickness, which may be estimated at 12,000 to 15,000 feet in North Wales.   The materials of which the lower members (the Arenig and Llandeilo beds) are composed were originally in the form of exceedingly fine mud, but towards the north-west of Scotland and of Ireland they are represented chiefly by sandy sediment (now converted into quartzites) giving evidence of the proximity of original land, the source of these materials, in those directions.   It is probable, therefore, that the same continent which produced the matter of which the Laurentian beds are composed was again laid under contribution, together with lands formed of these latter materials themselves; and that great rivers entering the sea from the west and north carried down fresh stores of sand, silt, and mud wherewith to overspread the floor of the ocean while it was undergoing gradual depression.   In this manner, it may be supposed, were formed the strata which afterwards were to be raised into the highest elevations of the British Islands and of Europe.*

The physical conditions of the Laurentian epoch over the region now under consideration were

---

* It is highly probable that the crystalline core of the Alps is composed of rocks of this age.   At Werfern, in the Saltzburg district, as stated by Murchison, strata with Silurian fossils have been detected, though they are generally metamorphosed.   (' Siluria,' 4th ed.)

repeated in that of the Lower Silurian; and thus far there is no evidence, as far as one can see, either of an Atlantic Ocean, or of a European continent.

(d.) *Upper Silurian and Devono-Silurian epochs.*— Over the region of the British Isles the Upper Silurian strata are physically disconnected from the Lower; these latter having been bent, folded, and even metamorphosed, then elevated in some places out of the waters of the ocean and exposed to denudation, ere the strata of the succeeding epoch began to be deposited. In this manner the spacious floor of the Lower Silurian sea was converted into a series of land ridges, gulfs, and bays,* while (as we may suppose) similar tracts of the Atlantic continent were submerged and converted into basins. The beds of the period of which we are now treating are found distributed at intervals over the British and European area, as in the cases of the Harz and "the Bohemian Basin,"† and only found at distant intervals over an area stretching from the centre of Europe to the western shores of Ireland. These latter tracts are of special interest in our inquiry, because they point unmistakably to the existence of extensive land areas in that portion of the present Atlantic immediately westward of the south of Ireland, and now overspread by waters of profound depth. Let us for a few moments consider the bearing of these western-most deposits on the question of

---

* Plate IV. and Expl., p. 66 *et seq.*

† Classic ground amongst geologists owing to the exhaustive labours of Joachim Barrande, as detailed in his 'Système Silurien de la Bohême.'

the existence of an Atlantic continent as late in time as the Upper Silurian and Devono-Silurian periods.[*]

It is well known that along the western coast of the Dingle promontory the Upper Silurian beds, charged with characteristic fossils of the Llandovery, Wenlock, and Ludlow stages, are to be seen passing upwards conformably, and with intercalations, into "the Dingle" or "Devono-Silurian" beds, the whole forming a continuous series of strata not less than 12,000 feet in vertical thickness. The upper beds of the series rise into the mountains and valleys of Kerry, and strike westward into the Atlantic Ocean. These beds were formed in a portion of a basin which comprehended not only the south of Ireland, but stretched across the south of Wales, England, and the centre of Europe. The land of the period, formed of Lower Silurian beds, lay to the north-wards, but of that to the south and west we have no trace; it lies submerged beneath the waters of the present Atlantic Ocean. Nevertheless, it must necessarily have been of great extent and prolonged existence, for the following reasons.

In the first place, it is probable that the Upper Silurian and Dingle beds were not by any means deep-water deposits, their nature and fossil contents forbid this supposition. They were probably deposited

---

[*] The Devono-Silurian beds lie on the confines of the two formations of which the name is a combination. They include the "Dingle beds" of Ireland, and the "Lower Old Red" of Scotland, as explained in the author's paper "On a proposed Devono-Silurian formation" (Quart. Journ. Geol. Soc., May 1882).

in a basin, which, as it gradually subsided, was kept pretty shallow by the influx of silt, sand, and mud, poured in at the mouths of rivers descending from the unsubmerged lands. But as the basin subsided its area must have been extended; while the land area, including ever-widening regions of the Atlantic, must, in order to remain above water, have slowly and continuously risen. The total amount of elevation on the Atlantic side equalled at least the total amount of depression on the British side, amounting approximately to 12,000 feet, as stated above. Considering, therefore, the continental condition of the Atlantic area in the preceding epoch, it is impossible to suppose that it could have been otherwise than, in the main, a land area over the tract west and south of the British Isles during the period of which we are now treating.

In the second place, the enormous mass of the Upper Silurian and Dingle, or Glengariff, beds, could only have been furnished by lands of large extent and of prolonged duration as such. The amount of these materials previous to denudation is the measure of the extent of wasted land ; and it is to be recollected that the original horizontal area was vastly greater than the present, not only on account of the denudation which has taken place over the south of Ireland, but also because the whole of this region has since been contracted in its horizontal dimensions by the lateral crushing of its strata, owing to which they are now sharply flexured and folded along axes ranging in nearly east and west directions. If these flexures were spread out the existing land would probably be three times as wide as at present.* On all these grounds we are driven to the conclusion that a large land area occupied that portion of the present Atlantic Ocean lying south of the 50th parallel of N. lat. and west of the 20th meridian W. long., a tract where the depth of the ocean bed is still comparatively moderate.† Similar beds recur in West Galway and Mayo on the Atlantic border.

By a similar train of reasoning the presence during the same period of extensive tracts of land in the region of the Atlantic lying to the westward and northward of the highlands of Scotland may be inferred. The great basins of " Lower Old Red Sandstone " (Devono-Silurian beds), lying along the eastern flanks of the north highlands, equally necessitate the contemporaneous existence of such lands as do the contemporaneous strata of the southwest of Ireland. These beds attain a thickness of over 16,000 feet,‡ and they were in all probability never deposited in deep water. As they were being spread out over a subsiding lake bed, the lands from which the sediment was derived would have been themselves submerged, unless they had experienced a somewhat proportionate uprising; while the vast

---

* Phys. Geol. and Geog. of Ireland, p. 130.

† Between 1000 and 1400 fathoms. The central Atlantic plateau, the maximum depth of which is about 2000 fathoms, and which on the western side rises to only 1240 fathoms, was probably the region of continental land in Palæozoic times. The deep depression, which, south of the banks of Newfoundland, reaches 3825 fathoms, was probably produced contemporaneously with the elevation of the Alleghanies (see chap. iv.).

‡ Prof. Geikie, Trans. Roy. Soc. Edin., vol. xxviii.

quantity of material brought down by the rivers of the period (of which the present deposits are only a fragment) involves the inference of proportionately extensive tracts beyond the margin of their basins. This tract is also still comparatively shallow.*

So far, then, as our evidence goes, we are brought to the conclusion that, down to the commencement of the Devonian epoch, the region of the Atlantic was mainly in the condition of land, while Europe and the British Isles consisted of a series of gulfs, sea-lochs, and inland lakes, separated by intervening ridges formed out of Laurentian, Cambrian, and Lower Silurian strata.

(e.) *The Carboniferous epoch in the British Isles.*— As bearing upon our inquiry it is scarcely necessary to dwell upon the physical conditions of the Devonian period†; we, therefore, pass on to those of the succeeding Carboniferous, which presents us with very interesting evidence regarding the relative positions of land and sea. In the first place we must consider the thickness of the sedimentary strata alone, attaining in the north of England, South Wales, Belgium, and Rhenish Prussia to 15,000 or 20,000 feet; so that the land necessary to supply so vast an amount of material (of which only fragments now remain) must have been itself of proportionately vast extent.‡ This land included the British highlands in part, but only a very small proportion of the sediment was yielded

---

* See Contour Map of the Atlantic in 'Voyage of the *Challenger*,' vol. ii. † For which see p. 78.

‡ Compare Figs. 1 and 2 of Plate VII. and Explan., p. 86.

by their diminutive surfaces. The main sources lay beyond and outside the northern and western uplands of Britain and Ireland; nor are we left in doubt regarding the direction in which these lands lay, as may be gathered from the following considerations.

If we draw a line from North-west Lancashire to the centre of England on the one hand, and again from South Wales to the centre of England, and protract the ascertained combined thicknesses of the strata (exclusive of the limestones) at various stages along these two lines, we shall find that the thickness diminishes in each direction towards the centre; or conversely, that the thickness of the strata increases as we proceed either to the north-west or to the south-west of England.* Here, then, we have indices clearly pointing in the directions of transport of material. On the principle I have already laid down, we infer that this swelling out of the sediments points to the position of the lands whence they were derived, namely, to the north-west and south-west of the British Islands.†

But the evidence does not end here, it is supplemented and confirmed by phenomena of an independent, though analogous, kind. In the above references to the comparative thicknesses of the

* The reader will find full details on this subject in the Quart. Jour. Geol. Soc., vol. xviii. pp. 127-146.

† My friend, the late Sir C. Lyell, fully admitted the cogency of this evidence for the existence of land in the Atlantic in Carboniferous times, and it affords me gratification to add something here in defence of those "Principles of Geology" which he illustrated in so masterly a manner.

Carboniferous strata, I have omitted those of the Carboniferous limestone, this member having had its origin in organic, not in mechanical, agencies. Nevertheless it is also capable of furnishing valuable evidence regarding the position of the Old Carboniferous lands.

The Carboniferous limestone of the British Isles occurs in its condition of greatest thickness and purity,* along a tract lying east and west from Galway Bay on the shores of the Atlantic across the centre of Ireland into Derbyshire. It also occurs in a similar condition of purity from the shore of the Bristol Channel into Belgium; as may be inferred from its character where it emerges from beneath newer formations. But if we observe this formation as we recede either northwards or southwards from those districts where the rock occurs in its condition of a pure limestone, we shall find that it undergoes marked changes in its composition consequent on changes in conditions of deposition. In the case of Ireland, the nearly solid limestone of the central parts,† 2500 feet in thickness, when traced into the extreme south-west of Cork and Kerry, thins away or passes into great beds of shale,‡ while in the north it passes into a formation of shale, sandstone, and conglomerate. In each case, the calcareous character gives place to the sedimentary. A similar change

---

* That is, freedom from beds of sandstone or shale.

† The middle division of the Carboniferous limestone of Ireland over the central area contains shaly beds, but in parts of Clare, Limerick, and Tipperary these can scarcely be recognised.

‡ As originally pointed out by the late Prof. Jukes.

is found over the British area. The solid limestone
of Bristol is represented in Devonshire by shales,
with thin impure bands of limestone, while that of
Derbyshire is represented in Scotland by sandstones,
shales, and a few limestone bands of no great
thickness.

Thus it will be evident, that while over the central
region of the British Islands, the sea waters of this

FIG. 1.—SKETCH-MAP OF THE BRITISH ISLES.
(Showing directions of unsubmerged areas in the Carboniferous period.)

period were clear, and almost free from sediment;
on the other hand, both to the north and south of
the central region they were laden with impurities.
From this we draw the inference that the lands
from which this sediment was derived must necessarily have lain in these directions; in other words,
from land lying north of the 55th and south of the
50th parallel of N. lat.

In the accompanying little sketch-map of the

British Isles I have shown by the arrow-headed lines the directions in which the sedimentary strata augment in thickness, as well as those in which the limestones give place to beds of sandstone and shale. It will be seen that they point in directions now occupied by deep waters of the ocean, but where in Carboniferous times lands of no small extent must have lain. Thus are we brought back to the conclusion, that as in the Upper Silurian and Lower Devonian, so also in the Carboniferous periods, the eastern section of the present ocean, even in its deeper portions, was in the condition of land, and that down to this epoch, at least, the Atlantic Ocean, as such, had no existence.*

(*f.*) *The Carboniferous epoch in North America.*— Nor are we without corroborative evidence of the correctness of our conclusion, if we look across to the other side of the Atlantic, and examine the arrangement of the Carboniferous strata over the continent of North America. We find in fact, *mutatis mutandis*, that the changes in the nature and thickness of the strata are precisely similar to those of the British Isles; only that in this case, the change from calcareous to sedimentary strata, and the augmentation of thickness, takes place in a north-easterly, instead of north-westerly or south-westerly direction.

* The bed of the ocean south-west of the British Isles descends to a depth of 2625 fathoms at a distance of about 400 miles from the Lizard, but afterwards becomes shallower. To the north-west of Scotland the ocean is of no great depth. The deep depression of the Bay of Biscay is probably contemporaneous with the post-Carboniferous upheaval of the Pyrenees.

Into the details of these phenomena it is unnecessary for me here to enter, as I have discussed them elsewhere.* Suffice it to say, that the Carboniferous limestone of the central States passes into, and is represented by, sedimentary strata in the north-eastern districts of the British possessions (Nova Scotia, &c.), while the sedimentary beds swell into vastly greater proportions in the same direction. These phenomena all point to the existence of large tracts of land in a north-easterly direction as regards North America, that is in the region of the present Atlantic Ocean, during Carboniferous times, and serve to confirm the conclusion I have arrived at, that the whole region of the North Atlantic was land, probably including Greenland of the present day, and was the source from which the strata both of Britain and North America were derived.†

The presence of a tract of ancient land to the north and west of the British Isles during the Carboniferous period is admitted by Mr. R. Godwin-Austen; and he also indicates the presence of old land both to the west and south of these Isles during the Middle and Lower Palæozoic periods; but he believes in the presence of a " great valley which, from the earliest times, has been the parting line between two sets of representative and synchronous

---

* The reader will find the subject fully dealt with in the author's paper "On Isodiametric Lines, &c.," published in the Quart. Journ. Geol. Soc., vol. xviii.

† The remarkable similarity of the vegetation of the Carboniferous period in America and Europe indicates a connection by land over which the plants migrated from one region to the other.

faunas." * But why, it may be asked, should an *ocean valley* be a line of division between two sets of *marine faunas?* Nay, is it not more probable that it was a *land ridge* which constituted this physical barrier; such a ridge, in fact, as I have inferred to have occupied the position of this ocean in Palæozoic times?

* Quart. Journ. Geol. Soc., vol. xii., p. 42.

CHAPTER VI.

THE GENESIS OF THE NORTH ATLANTIC OCEAN.

I DATE the genesis of the North Atlantic Ocean, properly so called, from the close of the Carboniferous period; and, consequently, from the same period, that of the British Isles and western Europe. My reasons are briefly as follows:—

(*a.*) *Formation of ridges and depressions at the close of the Carboniferous period in Britain, &c.*— Upon the close of the Carboniferous epoch, which was one of prolonged repose and general subsidence over the region of the British Islands and western Europe, the contractile forces of the earth's crust came powerfully into operation; large tracts previously submerged were elevated into land and mountain ridges, owing to the lateral pressure exerted along certain lines, by which the strata were forced high into the air, and were crushed, folded, and contorted.* That remarkable ridge, since greatly denuded, and partially concealed by newer formations, consisting in the main of contorted and

---

* Compare Plates VII. and VIII., figs. 2, with each other, in order to see the relative areas of land and water during the Upper Carboniferous and Permian epochs in the British Islands.

folded Carboniferous and Devonian rocks, which ranges from the shores of the Atlantic at the extreme south-west of Ireland through the south of England below the Cretaceous beds of the Thames Valley, and, reappearing in France and Belgium can be followed eastwards beyond the Rhine, belongs to this epoch, or in other words is of post-Carboniferous age. The highlands of Scotland and of the north of Ireland, previously almost smothered under Carboniferous strata, were at this time again upraised and brought to light, and this movement was accompanied by the formation of many minor ridges over the northern parts of our islands. That the Alps,* the Pyrenees, and the regions bordering them experienced the force of tangential pressure, and were subjected to one of their oft-repeated evolutions at this epoch can scarcely be doubted. It will be observed that all the ridges here described have a somewhat easterly and westerly trend, and are cases which may be quoted in support of Leopold von Buch's grand hypothesis.† Tracing these flexures and ridges westward, we take leave of them on the shores of the Atlantic, but they pass below its waters to unknown distances. Along with the ridges were formed furrows as deep as the ridges were high; and these carried westwards may be supposed, without much demand upon our credulity, to have formed the earliest abysses of the Atlantic. To this epoch probably belongs the deep depression which follows

---

* Carboniferous beds are found high up amongst the mountains of Savoy.

† The parallelism of contemporaneous mountain chains.

the 45th parallel north-westward from the Bay of Biscay into mid-ocean.*

Taken by themselves, the above arguments might not be considered to prove more than that the region of the Atlantic Ocean was converted into a series of ridges and hollows; but it is to be recollected that during the epoch we are now speaking of there was a general upraising of the British and European area, and we may therefore assume there was a corresponding subsidence of the adjoining region of the Atlantic. But I have reserved to the last what I consider a still more cogent argument, drawn from the. physical relations of the western half of the ocean to the adjoining shores of the American continent.

(b.) *Ridges and depressions in the Western Hemisphere.*—The boundaries of some of the oceanic regions of the world run parallel with ridges of mountains of the adjoining mainland, and no geographer can doubt the physical connection between such lines of depression and of elevation. Of such lines we have a remarkable example in the case of the coast line of the western Pacific and the elevation of the Andes and Rocky Mountains. Nor can there be a question of the close physical connection between the Alleghanies of eastern America and the coast line of the North Atlantic. Let us inquire how this bears on the question of the age of the Atlantic Ocean.

---

* The coal-formation of the Asturias is highly inclined, the strike of the strata being nearly magnetic north and south (N. 25° W.), as I am informed by Professor O'Reilly.

(*c.*) *The Alleghanies, their structure.*—The structure of this remarkable range is now well understood, chiefly through the labours of the late Professor H. P. Rogers.* It has been popularly brought before British geologists by Lyell,† who has given a generalised section drawn across the range, illustrative of its structure, here reproduced.

The greater part of these mountains is formed of Carboniferous strata thrown into a series of grand foldings which range generally parallel to the axis of the mountain and the line of the Atlantic coast, that is E.N.E. and W.S.W. These foldings become more and more intense towards the eastern flanks, amounting ultimately to actual inversion of the strata; but on the western flanks they gradually lessen in intensity, passing into slight undulations, and ultimately, towards the central regions of America, resolving themselves into the widest and flattest of ridges and furrows, where the strata are but slightly removed from the horizontal position.

This plicated structure is clearly due to powerful lateral pressure—the outcome of tangential forces— acting on the side next the Atlantic seaboard, and in a direction perpendicular to the axes of flexure themselves. The fact that the foldings of the strata become less deep, and ultimately well-nigh disappear in the direction of central America (the faint vibrations, as it were, of the mighty shock whose focus lay far to the east) is a clear proof that the forces have acted only from one side. The general result

* 'Geol. of Pennsylvania.'   † 'Elements of Geology,' 5th edit., p. 392.

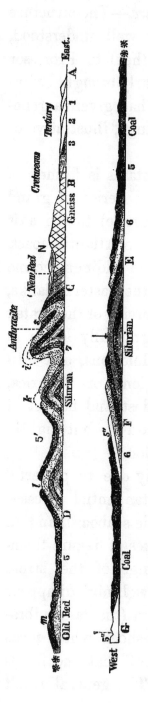

FIG. 2.—DIAGRAM EXPLANATORY OF THE GEOLOGICAL STRUCTURE OF PART OF THE UNITED STATES BETWEEN THE ATLANTIC AND MISSISSIPPI. Length from east to west 850 miles.

A B. Atlantic plain.
B C. Atlantic slope.
C D. Alleghanies, or Appalachian chain.
D E. Appalachian coal-field west of the mountains.
E F. Dome-shaped out-crop of strata on Ohio older than the coal.

F G. Illinois coal-field.
N. Falls and rapids of the rivers at the junction of the hypogene and newer formations.
i, k, l, m, Parallel folds of the Appalachians becoming successively more open and flatter in going from east to west.

References to the different Formations.

1. New Tertiary.
2. Eocene Tertiary.
3. Cretaceous strata.
N. Red Sandstone, with ornithichnites (New Red or Trias), usually much invaded by trap.
5. Coal-measures 1 (thus coal).
5'. Anthracitic edge measures.
5" Carboniferous line of the Illinois coal-field, wanting in the Appalachian.

6. Old Red or Devonian, olive slate, &c.
7. Primary fossiliferous or Silurian strata.
B. Hypogene to, or gneiss, mica shist, &c., with granite veins. Archæan or Laurentian (?)
Note.—The dotted lines at i and k : press tions of rock ved by denudation, the amount of which may be used by supposing similar lines prolonged from other points where different strata end abruptly at the side.
N.B.—The lower section at ** joins on to the upper one at **.

appears to have been that the Alleghany chain was thrust up into a great, crenellated, eccentric arch; the axis or fulcrum of which lay somewhere inland along the Atlantic slopes, while a corresponding depression (as we may confidently infer) was produced along a parallel tract now forming the bed of the ocean and the channel of the Gulf Stream, and probably bringing down with it a still wider tract of less disturbed strata stretching away under the centre of the Atlantic;—just as the uprising of the Alleghanies brought up with it the land region of Central America formed of nearly horizontal strata. Here then is a proposition which I commend to the consideration of physical geologists and geometricians. I put to them the question;—is it not in the last degree probable that while eastern America was thrust upwards, its strata crushed together and forced in upon themselves by irresistible lateral pressure, there was also produced a corresponding parallel depression situated to the eastward under the present bed of the ocean, in which we may suppose there is a sort of repetition, but in an opposite direction (downwards instead of upwards) of the folds and flexures visible to us on the land side? Such at least is the proposition I venture to submit; a thesis for the explanation of the Unknown based upon the Known.

It now only remains to show that the geological date of the depression on the American side corresponds to that of the depressions on the British and European side, which, as I have shown, are chiefly

E

referable to the period immediately succeeding the Carboniferous.

As Carlyle observes, that "in every phenomenon the beginning must always remain the most notable moment,"[*] it is important for our inquiry that we should endeavour to fix the date (geologically speaking) of this beginning of the Atlantic Ocean, and consequently for western Europe and eastern America; and, as we shall see, this important "moment" was synchronous on each side of the present ocean.

(*d.*) *Geological date of formation.*—The general parallelism of the chain of the Alleghanies to those of post-Carboniferous age over the British and European area is a phenomenon of great interest, and suggests the idea of identity of geological age; this, however, is fortunately not left to surmise only, as we can adduce direct evidence tending in the same direction. The evidence is briefly as follows :—

Lying along the eastern base of the Alleghanies, and reposing on the contorted and denuded edges of the Palæozoic rocks, occurs a small tract of coal-bearing strata called "the Richmond coal-field," formerly supposed to be of Jurassic age, but more recently determined to be of Triassic.[†] The relations of these strata to those on which they rest sufficiently show that the upheaval and denudation of the Alleghanies was anterior to (at least) the Triassic epoch; while, as the strata out of which the

---

[*] 'Sartor Resartus.'
[†] On the authority of Dr. O. Heer and M. J. Marcou.

plications have been formed is chiefly Carboniferous, the disturbances are subsequent to this epoch. This brings us sufficiently near the date of the British and European terrestrial movements for our purpose; because, although the more powerful flexurings were those which preceded the Permian epoch, they were continued until the commencement of the Triassic.*

Recapitulation.—Let me now briefly recapitulate the foregoing theses and inferences regarding the changes in the wide region of the northern hemisphere, stretching from central America on the one hand to central Europe on the other, through one hundred degrees of longitude, and extending as far south as at least the 35th parallel of latitude.

During the Upper Carboniferous epoch there seems to have existed a vast continental area, stretching southwards from the Arctic regions, embracing Greenland or its representative, and occupying the North Atlantic of the present day. The boundaries of this continent, east and west, approached the shores of the British Isles and of western Europe on the one side, and of the North American continent on the other. Now the conditions of these latter regions at this time were remarkable;—in fact unique, in the world's history.

* The lapse of time between the close of the Carboniferous and commencement of the Permian epoch, as represented by the deposition of strata, was very prolonged, and in the north of England was sufficient for the denudation of thousands of feet of Carboniferous strata, owing to which, in Lancashire and Yorkshire, the Permian strata are found resting on the Lower or Middle Carboniferous beds, the Upper having been denuded away in the interval. See 'Relative Ages of the Principal Physical Features of the Carboniferous District of Lancashire, &c.' (Quart. Journ. Geol. Soc., vol. xxiv., 323).

Neither before nor since have they been repeated on such a scale, and with such results; let us inquire what they were? The physical conditions under which the coal-fields of the British, European, and American tracts were formed have often been discussed and described. In all these countries they appear to have been as nearly as possible identical, and therefore the description, here necessarily brief, which is given for one part applies to the others.

We have reason to believe that the "Coal-measures" were accumulated over wide tracts of lagoons, or but slightly submerged basins, interrupted here and there by ridges and islands of older formations, but in the main stretching over vast regions as far as the margins of the continental lands which, as I have stated, occupied the region of the North Atlantic, and probably extended eastward and westward to embrace portions of Scandinavia and Canada. To the west and to the east this old Atlantic land formed a connecting ridge over which the vegetation of the period was able to spread from side to side, and from the margin of which it crept outwards into the marshy tracts and swamps when the conditions of depth and soil were favourable. These depressed, wide-spread regions were in the main, and generally speaking, at the level of the ocean—sometimes a little lower, in which case the sea waters flowed in and inundated considerable tracts, at other times a little higher; but the whole area must have been gradually sinking, and as gradually replenished with sediment brought down from the central continent by numerous large rivers. During pauses in the subsidence, and when the conditions were favourable, the plants and forest trees of the Carboniferous period took possession of the marshes and lagoons, their roots and stems partially submerged, and by their decay forming the vast accumulations of vegetable matter which have subsequently been converted into beds of coal. The result of this long uninterrupted process of slow accumulation of beds of sand, mud, silt, and vegetable matter was the formation of the Coal-measures to a depth of several thousand feet vertical. But this extended epoch of slow waste on the one hand, and of quiet accumulation on the other, was destined to be brought to a decisive close; when, at the termination of what we call the Carboniferous period, the resistless contractile forces of the earth's crust came into operation, resulting in the uprising of the Alleghanies and of central America on the one side, and of the ridges and plains of the British Isles and of western Europe on the other. Along with these uprisings, and the crushing together of the strata, came, as I have already inferred, corresponding depressions over the region now forming the North Atlantic Ocean. The waters previously banked out, or only gaining occasional access, for evermore flowed in and overspread these tracts, which became themselves more deeply submerged as the mountain ridges and adjoining slopes uprose into higher levels. In this manner, not only large areas of the recently formed strata became permanently submerged, but a great part at least of the continental land itself may be supposed to have gone down under the flood of waters. Henceforth in geological history, the North Atlantic, before a continent, now becomes an ocean. Instead of a source of sediment, a sort of quarry from whence the materials of continents were hewn, it now becomes a gulf of vast dimensions, into which has been poured through long ages the used up materials of these very continents themselves.

PART II.

PALÆO-GEOGRAPHICAL AND GEOLOGICAL MAPS OF THE BRITISH ISLES.

CHAPTER I.

(PLATE I., FIGS. 1 AND 2.)

THE LAURENTIAN PERIOD.

IT is intended to show by the Plate I., Fig. 1, those tracts where the Laurentian rocks reach the surface, and those under which they may be supposed to extend, though concealed beneath more recent formations; also (Fig. 2), the portions of the surface over the western part of the European area, including the British Isles, occupied by the land and sea of the Laurentian, or Archæan, period.

(*a.*) *Laurentian areas* (Fig. 1).—The Laurentian rocks appear at the surface (over the area embraced by the map), as forming the greater portion of the Scandinavian promontory,* in the north-western highlands of Scotland,† and the outer Hebrides, in the north-west of Ireland and Galway,‡ in the centre

---

* " The Tellemarken formation," of Kjerulf and T. Dahll.

† As first recognised by Sir R. I. Murchison (1858–59), Quart. Journ. Geol. Soc., vol. xv. p. 354, and vol. xvi. p. 215.

‡ Rep. Brit. Assoc. York (1881), and Trans. Roy. Dub. Soc., vol. i. n.s. (1882). Some geologists believe in the presence of Pre-Cambrian beds in Anglesea. As an officer of the Government Survey I reserve my opinion.

and north-west of France,* in the Province of Madrid in Spain,† and along the margin of the Silurian basin of Bohemia. They may be supposed to underlie all the remaining portions of the land, except those districts formed of intrusive granitic or trappean rocks, which, as compared with the former areas, are very small, and can only occasionally be represented on a map of the scale here adopted.

(*b.*) *Nature of the Laurentian rocks.*—These rocks consist of foliated granite or gneiss, hornblendic and micaceous schists, crystalline limestone or marble. The gneiss is generally massive, porphyritic, and of a red colour, consisting of orthoclase, oligoclase, quartz, and mica of two varieties.‡ The thickness of these beds in Canada has been estimated by Sir W. Logan § at 35,000 to 40,000 feet. In Britain they are probably not much less in vertical dimensions, only in horizontal extent.

The Laurentian rocks have undergone intense metamorphism, owing to which they now only occur in a crystalline condition. Originally, there is every reason to believe, they were formed of sedimentary materials, such as those of the Lower Silurian system, consisting of sandstones or grits, slates, flagstones,

---

* This is inferential.

† Gneiss pierced by granite, surmounted by schists and sandstones of Lower Silurian age, as recognised by Casiano de Prado and De Verneuil.

‡ To this formation the red granite of the Nile valley and the borders of the Red Sea probably belongs. Lartet, "La Geologie de la Terre Sainte," in 'Voyage d'exploration de la Mer Morte,' par M. Le Duc de Luynes, vol. iii.

§ 'Geology of Canada,' p. 45.

and limestones, all of marine origin; and whether or not the *Eozoon Canadense*, found in this formation in Canada, be a true organism, the occurrence of beds of limestone leads us to infer that the ocean waters of this early period of geological history were not destitute of living creatures, though probably of very simple organisation.

(*c.*) *The Laurentian continent (Atlantis).*—As the Laurentian rocks form extensive tracts both in North America and Europe, it may be inferred that the land which was the source of the sediment of which they are composed was situated in a region lying between these two areas; that is to say, in the region of the Atlantic Ocean, including probably the continent of Greenland, and possibly the Polar regions. It may be supposed that large rivers flowed down into the ocean of the period, both towards the west and towards the east, and that this sediment was deposited over the floor of the Laurentian ocean, now occupied by North America and Europe. The margins of this land are necessarily only approximately inferential.

CHAPTER II.

(PLATE II., FIGS. 1 AND 2.)

THE CAMBRIAN PERIOD.

THE physical conditions of the Cambrian period over the British area contrast strongly with those of the Laurentian.   Previous to the deposition of the Cambrian beds,* those of the preceding Laurentian period had been metamorphosed, elevated into land areas, and largely denuded, so that the bed of the Laurentian ocean (Plate I., Fig. 2) now appears as part of a large continental area, embracing the northern and western portions of the British Isles; while the ocean extended over the western districts of Europe, the whole of England, and parts of Scotland and Ireland.   I have in previous pages already touched at some length on this part of the subject.

(a.) *Submerged Cambrian areas.*—From considerations stated at length elsewhere,† I have arrived at the conclusion, that during the Cambrian period, an

---

* I use the term Cambrian to include the Longmynd, Harlech, and Llanberis beds, together with the overlying Upper Cambrian Lingula flags, as the fauna of the latter has been shown by Dr. Hicks to be present in the former. I therefore take the base of the Silurian series at the Tremadoc slates.

† Quart. Journ. Geol. Soc., May 1882, p. 210, and Brit. Assoc. Rep. 1881, p. 642.

Archæan ridge formed of Laurentian strata stretched through the British Isles in a S.W. and N.E. direction, embracing the region of the west of Ireland, and of the Grampian mountains, by which the Cambrian beds of the N.W. highlands of Scotland were separated off from their representatives of the English and Welsh areas. This conclusion depends in part on the extreme dissimilarity existing between the representative beds on either side of the supposed ridge. I shall, therefore, describe the beds under the two types which I have called, on the occasion referred to, those of the " Caledonian," and of the " Hiberno-Cambrian."

(b.) *Cambrian beds of the Caledonian type.*—These are restricted to the north-western highlands of Scotland, where they occur interposed between the Laurentian rocks below, and the quartzites, limestones, and shales of the Lower Silurian age above. They consist of great beds of red and purple sandstone and conglomerate, generally in nearly horizontal positions, forming bold escarpments, and isolated pyramidal masses. The pebbles of which they are mainly formed consist of various kinds of gneiss, schist, porphyry, and quartzite ;—presumably derived from the adjoining land areas of Laurentian strata. Sir A. C. Ramsay considers these beds to have been deposited in the waters of an inland lake, of which the outer Hebrides formed the western margin.* In this view I concur. No fossils have been found in these lacustrine beds.

* Phys. Geology and Geography of Great Britain, 5th ed., 283, &c.

(c.) *Cambrian beds of the Hiberno - Cambrian type.*—These are vastly more extensive than the former; and, though they only crop out to the surface in a few places, may be presumed to underlie nearly the whole of England and Wales, as well as the adjoining parts of Europe. They consist of green and purple massive grits, quartz rocks and slates, with, occasionally, pebbly beds; and as the fauna is distinctly marine the beds may be inferred to be of oceanic origin. Cambrian rocks of this type are found in the east of Ireland, with *Oldhamia*, a sertularian zoophyte, and annelid tracks and borings, such as those of *Histioderma*. In North Wales and Shropshire they have yielded trilobites,* while the Upper Cambrian beds are rich in marine forms. At St. David's, Dr. Hicks has brought to light several genera of trilobites in beds contemporaneous with those of the Longmynd and Harlech group of North Wales. These rocks also are found in Charnwood Forest, in Leicestershire,† where they are in some places altered or metamorphosed, and associated with trap rocks; in the Ardennes mountains, on the borders of France and Belgium, where they consist of quartzites, quartz-schists, and schists, with *Oldhamia radiata* (one of the Irish species), *Dictyonema sociale*, *Lingula*, and tubes or impressions of annelids.‡ In the Système Salmien, forming the

---

* Discovered by Mr. Salter in the Longmynd beds.

† I regret I cannot agree with Dr. Hicks and several other distinguished geologists in regarding these beds otherwise than of Cambrian age, to which they were originally referred by Professor Jukes.

‡ Dr. Mourlon, 'Géologie de la Belgique,' t. 1, p. 31.

upper division of the series, trilobites of the genus *Paradoxides* have been discovered by M. Malaise.

These rocks also occur in Normandy and Brittany, consisting of green slates and grits, resting on gneiss and schist, which is probably of Laurentian age.[*]

Regarded as a whole, the Cambrian beds of the region now described are clearly of marine origin, and present—both in lithological characters, and from the occurrence of a marine fauna—a marked dissimilarity to the beds of the Caledonian type.

(*d.*) *Distribution of land and sea.*—Having already touched on this subject,[†] it may suffice here to state that during the Cambrian period the area of the North Atlantic Ocean consisted (at least in its eastern part), in the main, of land, but probably disposed in ridges and hollows trending in directions parallel to that of the Outer Hebrides, and of the Archæan ridge which I have described as traversing the British Islands. Between these ridges, lakes or arms of the sea may have been interposed—but to the east and south of the British Archæan ridge we may conclude, from the evidence already adduced,[‡] that the waters of the ocean spread for great distances over Europe, bounded perhaps to the northward by the Scandinavian ridge, and possibly broken by an occasional island of the older rocks in the region of central Europe. It is probable, however, that the waters were nowhere of any great depth.

---

[*] Dalimier, Bull. Soc. Géol. France, 2 ser., vol. xx.

[†] Page 30. [‡] Page 60.

CHAPTER III.

(PLATE III., FIGS. 1 AND 2.)

THE LOWER SILURIAN PERIOD.

WITH the commencement of the Lower Silurian period,* the ocean resumed the dominion it had partially lost during the preceding Cambrian period; and as time went on, the entire area of the British Islands and adjoining parts of Europe became submerged and covered with sediment. It may be confidently affirmed, that there is not a square mile over this region which was not originally buried beneath strata belonging to the Lower Silurian period. The old Archæan ridge was covered by strata still in existence; and even the Cambrian and Laurentian rocks of the north highlands of Scotland were, in the opinion of Sir A. C. Ramsay, submerged and buried under the accumulating piles of these strata before that era passed away.† Land was, however, probably not far away, and its position was to the north-west of the British Isles.‡

---

* I assume the base of the Lower Silurian series to be the Tremadoc slate, or (in their absence) the Arenig beds forming the lower part of the Llandeilo group.

† Phys. Geog. and Geol. of Great Britain, 5th ed., p. 87.

‡ I have already adduced evidence for this statement (p. 32).

(*a.*) *Nature of the Lower Silurian beds.* — The strata of this period consist of dark and grey slates, grits sometimes calcareous, and, rarely, bands of limestone. The fossils are all of marine genera. Over the northern and western areas of the British Isles these strata have undergone extensive metamorphism, so that in the highlands of Scotland, and of the north and west of Ireland, they consist of quartzites, micaceous, talcose, and chloritic schists, and crystalline limestones, sometimes serpentinous, with their varieties; presenting a marked contrast to their unaltered representatives in the south of Scotland, in Wales, and in the east of Ireland.

(*b.*) *Relations to adjoining formations.* —The Lower Silurian rocks are discordantly superimposed upon all formations older than themselves. This is the case in North Wales, in the east of Ireland, and in the north of Scotland. Owing to this discordance, and the large amount of denudation to which the Upper and Lower Cambrian beds were subjected at the close of the Cambrian period, we find the Lower Silurian beds resting on strata of various stratigraphical positions. Thus in North Wales and Salop, the Arenig beds are found resting sometimes (as near Bangor and Carnarvon) on the purple slates and conglomerates of the Cambrian series; * sometimes, as in Pembrokeshire and Merionethshire, on the Tremadoc slates. In the east of Ireland, in county Wicklow, the Lower Silurian beds rest discordantly on the Lower Cambrian beds; and in the

* Ramsay, *supra cit.*, p. 78.

north of Scotland, the quartzites and limestones, representing the Llandeilo beds, rest discordantly, sometimes on the Lower Cambrian beds, at others, on the Laurentian.

(c.) *Lower Silurian areas.*—The principal districts where the Lower Silurian rocks form the surface, are the north and central highlands and the southern uplands of Scotland, the lake district of the north of England, the north and centre of Wales, the north-west, north-east, and south-east of Ireland and the Isle of Man.   They also occupy portions of Normandy and Brittany,* where they rest on Cambrian and Laurentian beds, and they have been proved by boring below the Tertiary and Cretaceous strata at Bruxelles (Brussels), Louvain, St. Tron, Menin, and Ostende in Belgium.   They also appear at the bottoms of the valleys between the Sambre and the Meuse, as determined by Prof. Gosselet ; † they probably underlie a large portion of the Paris basin, where they are concealed by Tertiary and Cretaceous formations.

(d.) *The formation of marine origin.*—The fossils yielded by the Lower Silurian rocks, whether in the north-west of France, in Belgium, in Wales, in Ireland, or in the north of Scotland,‡ all go to prove the marine origin of the strata themselves. They consist chiefly of trilobites, molluscs—cephalopods, gasteropods (lamellibranchs not plentiful),

---

* Murchison, ' Siluria,' 4th ed., p. 408.

† Quoted by Dr. Mourlon, Géol. de la Belgique, p. 40.

‡ From the Durness, or Assynt, limestone.

and brachiopods—a few corals, and graptolites. The abundance of these forms in the calcareous beds proves that the waters of the sea teemed with living forms. In the Bohemian basin, as M. Barrande * has shown, there was a prodigious development of life; but often, through many hundreds of feet of slates and grits in some districts, no trace of organic structure is discoverable.

(e.) *Distribution of land and sea.*—From what has been said it will be understood that generally speaking the eastern portion of the North Atlantic area consisted of land from which the sediments forming the strata of the Lower Silurian age were derived. The limits of this land area and its margin cannot be even approximately determined, and on the map (Plate III., Fig. 2) have been only vaguely represented. From the margin, however, thus indicated, whether far from the present coast of the British Isles, or otherwise, the Lower Silurian ocean spread indefinitely southwards and eastwards. Its depth may have been in some parts—and especially during the Llandeilo stage—considerable, though probably not profound; towards the close of the period the sea became shallower through the accumulation of sediment and the upraising of the bottom.

* Syst. Sil. de la Bohême.

CHAPTER IV.

(PLATE IV., FIGS. 1 AND 2.)

UPPER SILURIAN AND DEVONO-SILURIAN PERIODS.

THE relative areas of land and sea during these periods differ widely from those of the period which preceded it, as will be seen on a comparison of Plates III. and IV.   The sea which overspread the whole of the British Isles and adjoining portions of France and Belgium in the Lower Silurian epoch, is now restricted mainly to the southern and central portions of the British Isles : while large tracts in the north and west, as well as Normandy and Brittany, are converted into land surfaces, holding in their deep depressions lakes or fresh-water basins, which were formed towards the close of the Silurian period; or in more definite terms, during the Devono-Silurian stage.

(*a.*) *Nature of the Upper Silurian beds.*—The basement beds of the Upper Silurian series (Llandovery beds) are frequently conglomerates and sandstones, derived from the disintegration of the rocks of the adjoining lands, and by their position we are able to indicate the position of the margins of these lands themselves.   Such is the case in the districts of

Connemara in the county Galway,[*] and of Builth in Radnorshire.[†] The succeeding beds consist of grits, shales and limestones of the Wenlock and Ludlow series, often rich in marine fossils. These beds occur in West Galway and Mayo, in North and South Wales, and Monmouthshire, in Staffordshire, along the southern slopes of the Cumberland mountains, and those of the southern uplands of Scotland. In the north of France and Belgium they are altogether wanting, as the Devonian beds rest against the shelving flanks of ancient lands formed of Lower Silurian and Cambrian strata.[‡]

(b.) *Devono-Silurian beds.* — Under this term I include a series of beds known by various names, and chiefly developed to the north and to the south of the British area. They include the " passage beds " of Murchison, and the " Downton sandstone," lying at the top of the Upper Ludlow rock, in South Wales; the " Dingle and Glengariff beds " of Jukes, forming the south-western mountains of Ireland, and seen to rest conformably on the Upper Silurian beds along the coast of Dingle; "the Fintona beds " of the north of Ireland, which rest unconformably on older crystalline strata; and the " Lower Old Red Sandstone " of Scotland. The Devono-Silurian beds form the connecting series between the Upper Silurian and the estuarine Devonian beds of Monmouth and South Wales, and are probably repre-

---

[*] Phys. Geol. of Ireland, p. 23.
[†] Ramsay, Phys. Geol. of England and Wales, p. 89.
[‡] Mourlon, Géol. de la Belgique, t. 1, p. 54.

sented south of the Bristol Channel by the "Fore-
land grits and slates" of North Devon.[*]

These beds were formed by accumulations, sometimes of great thickness, of green, red, and purple
sandstones, grits, shales, or slates, and conglomerates
—of marine origin in the southern portion of the
British area, but in the northern, probably of lacns-
trine origin.   In the latter district, according to the
views of Professor A. Geikie, the beds. of this division
of the series were deposited in several distinct lake-
basins.[†] One ("L. Orcadie"), north of the Grampians ;
a second (" L. Caledonia "), south of these mountains ; a third (" L. of Lorne "), a district north of
Argyleshire, lying at the entrance of the Great
Glen ; and a fourth (" L. Cheviot "), on the
southern borders of Scotland. [‡]   These deposits were
derived from the waste of the adjoining lands formed
of the metamorphosed beds of the highland mountains, but how far they extended in the direction of
the Scandinavian promontory is altogether uncertain,
so that the eastern limits of these basins have been
left on the map undefined.

(c.) *Relations to the adjoining formations.*—Throughout the British area the Upper Silurian beds are

* "On a proposed Devono-Silurian Formation," Quart. Journ. Geol.
Soc., May 1882, p. 200; also Trans. Roy. Dub. Soc., vol. i. p. 147
(1880).

† "On the Old Red Sandstone of Western Europe," Part I., Trans.
Roy. Soc. Edin., vol. xxviii. The margins of these basins drawn on the
maps (Plate IV., Fig. 2), are very much those indicated by Prof. Geikie.
He considers that Lakes "Orcadie " and " Caledonia " were never united.

‡ It seems probable that in its earlier condition this lake was connected
with the sea, but was subsequently disconnected.

unconformable to the Lower Silurian, and in some cases, as in Shropshire, they rest directly on Cambrian beds. In a word, the physical hiatus between the upper and lower divisions of the great Silurian system of Murchison is as marked and complete as it is possible to conceive between any two adjoining sets of strata; and this being the case, it is not to be wondered at that Sedgwick claimed as "Cambrian" all the beds below the Llandovery horizon. After the close of the Lower Silurian epoch, represented by the "Bala beds," there occurred, over the region in question, terrestrial disturbances of great intensity, accompanied in the north of Ireland and Scotland by metamorphic action.* Large tracts of the ocean bed were converted into land surfaces, while denudation ensued on a great scale, owing to which the uppermost Lower Silurian beds were washed away; and on the resubmergence of the depressed tracts, these materials were used up in the construction of the basement beds of the succeeding Upper Silurian series. Depression then went on, during which the Wenlock and Ludlow beds were formed under tranquil waters; and towards the close of the latter period the great lakes of Scotland and of the north of Ireland, bounded on all sides by metamorphosed strata, were formed; while vast masses of material

* That this metamorphism of the Lower Silurian beds of the North British area took place before the Upper Silurian period was first pointed out by Harkness in his paper "On the Age of the Rocks of West Galway," &c. (Quart. Journ. Geol. Soc., vol. xxii.), and has more recently been indicated by myself in the 'Phys. Geol. and Geog. of Ireland,' p. 22 (1878).

were accumulated over the region of the south-west of Ireland, but in this instance, probably, under the waters of the ocean.

(*d.*) *Upper Silurian areas.*—The principal areas of this series are to be found along the eastern borders of Wales, extending from the north coast at Conway southwards through Montgomeryshire, Shropshire, Radnor, into Hereford and Monmouth. Isolated portions rise from below the South Staffordshire coalfield, as at Dudley. Eastwards these beds extend under the Cretaceous rocks, and have been proved by borings at Ware,* in Hertfordshire, and no doubt, they extend eastwards to the coast. But in Belgium the Upper Silurian rocks are unrepresented, and the Devonian rocks lie in a trough, having the Lower Silurian beds of Brabant on the north, and the Cambrian beds of the Ardennes on the south, against the flanks of which the more recent strata were originally deposited.† As the "Devono-Silurian" beds are in all probability represented by the "Système Gedinnien" (in part at least) at the base of the Devonian series, the sea area in Plate IV., Fig. 1, is extended over the north of France and Belgium, along the line of this old trough.

The Upper Silurian beds lie along the southern flanks of the Cumberland mountains, having a southerly dip, and probably extend eastwards under

* Mr. R. Etheridge, F.R.S., the *Times*, 19th May, 1879; also, Presidential Address to the Geol. Society of London, 1881.

† Dr. Mourlon, Géol. de la Belgique, t. 1, p. 54.

the Carboniferous rocks to the coast. They again occur along the flanks of the southern uplands of Scotland, where towards the close of that epoch they were probably separated from the sea, and the area converted·into a lake, in which were deposited the beds of the Devono-Silurian series (Lower Old Red Sandstone) near St. Abb's Head.

In Ireland, Upper Silurian beds occur in Dingle (Kerry), passing upwards into the Devono-Silurian, or Dingle, beds;* they also occupy considerable tracts on both sides of Killary Harbour and the shores of L. Mask; and they are again found forming a small tract on the borders of Roscommon and Sligo. The areas of the Devono-Silurian beds have already been stated.

(e.) *Distribution of land and sea.*—As will be seen by referring to the map (Fig. 2, Plate IV.), the land areas of the region now under description lay both towards the north and towards the south, between which there was a gulf of moderate depth extending over the region of the south of Ireland, England, and the north of France, under which the marine strata were deposited. This gulf threw out an arm towards the north, but how far it stretched beyond the eastern coast line it is impossible to say. Much of the Atlantic area was also in the condition of dry land, as I have already explained.† The land area of the north was probably a prolongation of the

---

* Expl. Mem. Geol. Survey, sheets 160 and 170.

† I have already discussed at some length the question of the position of the Atlantic land area during this epoch, see *ante* p. 33 *et seq.*, so that it is unnecessary to go over this ground here.

Scandinavian promontory, while large tracts of the Atlantic to the westward formed continuous land with the northern highlands of Scotland and Ireland. The western and northern limits of this land area are incapable of definition; it may have included isolated basins besides those we are able to identify in North Britain. At the commencement of the period we are now dealing with, land prevailed to a much greater extent than that shown in the map, but as time went on, the areas of the sea and inland lakes were extended down to the commencement of the Devonian period.

CHAPTER V.

(PLATE V., FIGS. 1 AND 2.)

THE LOWER AND MIDDLE DEVONIAN PERIODS.

THE epoch represented in Figs. 1 and 2, Plate V. is that ranging through the Lower and Middle Devonian stages, embracing the beds of the "Lynton," "Hangman," "Ilfracombe," and "Morthoe" divisions of Devonshire, and those lying between the Système Gedinnien, and the Calcaire de Frasne of Belgium. This series, several thousand feet in thickness, is entirely marine. It is laid open in North and South Devon in England, passes below the Cretaceous rocks of the Thames valley, and reappears in numerous sections along the river valleys of Belgium, such as those of the Sambre, the Meuse, and the Ourthe, as well as along the valley of the Rhine and its tributaries. In a somewhat altered form it occupies a large tract of country bordering the valleys of the Usk and the Wye, in Monmouthshire and Herefordshire, and is generally, but erroneously as I believe, called by the name of "Old Red Sandstone." These last-named beds, I consider to have been deposited in an estuary, bounded towards the north-west and north-east by Silurian lands, but opening southwards into the sea, in which the Devonian beds were being contemporaneously formed. I have, therefore, called these beds "estuarine Devonian." * From the remainder of the British Islands, including the whole of Ireland and Scotland, the Devonian beds are absent, owing to causes which I shall presently endeavour to explain.†

(a.) *Nature of the Devonian beds.*—From what has been said, it will be inferred that the Devonian beds south of the Severn differ in some characters from their representatives north of that river.

South of the Severn the formation consists of beds of grit, shale, and limestone, in alternating masses, highly fossiliferous, and yielding remains of molluscs, corals, and crinoids, and some plants.‡ North of the Severn the beds consist of red and grey marls, with earthy calcareous bands ("cornstones"), and red or purple sandstones. Fish remains are present in the cornstones, but some examples of *Lingula* in the lower beds, and of *Serpula* in the upper, are all the evidences of invertebrate life which, up to the present, they have presented to us.

In South Devon the limestones are more massive, but it is not till we examine the sections in the

---

. * "On the Relations of the Rocks of the South of Ireland to those of North Devon, &c."—Quart. Journ. Geol. Soc., May 1880, p. 268. The term used in this paper is "lacustrine." I have since preferred the term *estuarine*.

† As I have shown in the paper above referred to.—*Ibid.*, pp. 264 and 270–3. Mr. Etheridge in his Presidential address expresses his concurrence in my views.—Quart. Journ. Geol. Soc., May 1881, p. 193 *et seq.*

‡ Mr. Etheridge has given a complete account of the fauna of Devonshire in his Presidential address, *supra cit.* He enumerates no less than 235 species as occurring in the Middle Devonian beds of South Devon.

Meuse and Ourthe, in Belgium, that we are able to appreciate the extent to which marine limestones were developed at this period.

(*b.*) *Relations to the adjoining formations.*—Confining our attention to the region of the south of England, the Devonian strata may be considered as forming a complete connecting series with the Upper Silurian and Devono-Silurian beds below and the Carboniferous beds above. Over this region, deposition of sediment appears to have proceeded with but few interruptions, of which none are marked by visible physical breaks. After the close of the Silurian period, depression went on, and various kinds of sediment were formed over the floor of the sea bed, during slow subsidence over this area. Meanwhile the fauna of the previous period, modified as regards species, but largely similar as regards genera, reappeared under new forms; and, as Mr. Lonsdale long ago observed, presents generally a *facies*, intermediate between that of the Carboniferous on the one hand, and of the Silurian on the other. Mr. Etheridge recognises about 550 species as belonging to the British Devonian group.

(*c.*) *Absence of Devonian beds in the north and west of the British Isles.*—The absence of representatives of the marine Devonian beds of the south of England over the Irish and Scottish areas is a circumstance which, in my opinion, can only be satisfactorily accounted for in one way, namely, that these areas had been elevated into dry land during the time that the south of England and adjoining continental regions were submerged beneath the waters of the Devonian sea, and became the receptacles of Devonian sediment.* As confirming this view we have the fact that the Upper Devonian, or Old Red Sandstone proper, is everywhere unconformable to the beds on which it rests in Ireland and Scotland, whether these belong to the Devono-Silurian or still older formations.† There is, therefore, in these countries a gap, or hiatus, of a very decided character, which is not the case in Devonshire, where the whole series, from the base of the Devonian into the Lower Carboniferous series, is complete. This northern and western hiatus is, in fact, filled up in Devonshire, owing to the presence of the Lower and Middle Devonian beds, which are absent in Ireland and Scotland.

The unconformity between the Upper Devonian Sandstone (or Upper Old Red Sandstone), and the Devono-Silurian beds (i. e. the "Dingle beds" of Ireland, and the "Lower Old Red Sandstone" of Scotland), indicates that after their deposition these latter were subjected to disturbances, elevated into land surfaces, and exposed to denudation. In this position they remained throughout the Lower and Middle Devonian periods, and were only resub-

---

* This view was first proposed in the 'Geological Magazine,' and was afterwards more fully unfolded in the paper above cited, and in the Trans. Roy. Dublin Soc., vol. i. p. 147, &c.

† This is distinctly enforced by Sir R. Griffith as regards Ireland, and is exemplified in many sections, especially those of the Dingle promontory ; and by Professor Geikie as regards Scotland. There may, also, be a slight unconformity at the base of the yellow sandstone in S. Wales, in keeping with that of the adjoining Irish area.

merged when that of the Upper Devonian set in. Plate V., Fig. 2, represents the period of elevation of the west and north of the British Isles, and of the concurrent depression of the region to the south.

It is probable, also, that the centre and north of France (Normandy, Brittany, and the Ardennes), were in a condition of land surfaces during the deposition of the Lower and Middle Devonian beds, as these everywhere rest, and with varying geological horizons, against the older formations of which this part of France is largely formed. We must also recollect that the whole area of the south of the British Isles, and of the adjoining parts of the continent, has undergone enormous lateral compression, in a north and south direction, owing to which the originally horizontal Devonian and Carboniferous beds have been crushed into numerous sharp foldings and flexures, lying along approximately east and west axes, and that these extend from the extremity of Kerry and Cork through Devonshire, under the Thames valley, and reappear in France and Belgium, and beyond the banks of the Rhine.

If, therefore, we wish to realise the geographical position of the Devonian beds as originally deposited, we must flatten out these flexures and reduce the beds to the horizontal position, in which case the present apparently narrow trough running across the south of England and north of France would be spread out to probably almost twice its present width.*

---

* The flexuring of these beds, as laid open along the Meuse, is very well shown by M. Gosselet in a drawing, as copied by Dr. Mourlon, in the

(*d.*) *Distribution of land and sea.*—On the above grounds, therefore, I have represented in Fig. 2, the whole of the western and northern portions of the British Islands, with the adjoining portions now covered by the ocean, as land during the Middle Devonian period. Contemporaneously with this the sea extended over the south of England, and eastwards into Germany, under the waters of which were deposited in England the fossiliferous limestones of Ilfracombe and Plymouth; in Belgium, the " Calcaire de Givet"; and in Germany, the "*Stringocephalus* limestone." Once we thoroughly understand the physical relations of these different areas, the reasons for the present distribution of strata become clear.

Geol. de la Belgique, t. 1, p. 56. As the average angle of inclination exceeds 45 degrees, the original length would have been in this case more than twice the present.

CHAPTER VI.

(PLATE VI., FIGS. 1 AND 2.)

OLD RED SANDSTONE AND LOWER CARBONIFEROUS PERIODS.

IN order to save the engraving of a separate plate, I have endeavoured to include the above groups of strata in one pair of maps, although the Old Red Sandstone is a member of the Upper Devonian series, rather than of the Carboniferous. This is proved by the occurrence of Old Red fishes (*Coccosteus, Pterichthys, Asterolepis,* &c.), together with plants (*Adiantites Hibernicus*), and fresh-water mollusca (*Anodonta Jukesii*), in the beds of this formation in Ireland. In many districts, however, the Old Red Sandstone appears to be conformable to the over-lying Lower Carboniferous beds, while unconform-able to all strata older than the Upper Devonian beds; and as these are only present in the south of England, Belgium, and France, the Old Red Sand-stone is elsewhere unconformable to the strata on which it reposes.

The strata included in Plate VI., range from the Old Red Sandstone or Conglomerate to the top of the Carboniferous limestone. At the commencement of the deposition of these beds the greater part of the area now described existed as land. But as time went on, the British area became depressed, and the sea gradually gained on the land; so that, at its close, only the northern and western tracts were unsubmerged, together with portions of the border districts of Scotland. The Cumberland mountains, and a tract ranging from North Wales, Shropshire, and the centre to the east of England was also unsubmerged.* Over the submerged areas the Lower Carboniferous strata were deposited; from the unsubmerged districts they are absent. Throughout South Staffordshire, parts of Salop, Leicestershire, and Warwickshire, the Upper and Middle Carboniferous beds rest directly on those of Silurian or Cambrian age.†

(*a.*) *Characters of the Old Red Sandstone.*—The Old Red Sandstone is found over the south of Ireland in the form of a massive conglomerate, rising into fine escarpments in the Commeragh and Dingle mountains, and passing upwards into finer red sandstones and beds of flagstone and shale. The uppermost beds, called the "Kiltorcan beds," contain fish remains, a fresh-water mussel (*Anodonta Jukesii*), and plants. They are, therefore, lacustrine deposits over the area of the south of Ireland, and mark the upper limit of the Old Red Sandstone formation.

In South Wales, along the northern margin of the

---

* It is possible that the sea may have spread between North Wales and the Wicklow mountains during this time.

† South of Halesowen, the Upper Silurian beds were penetrated by a coal shaft under the Upper Coal-measures, and at Dudley, Forest of Wyre, Shrewsbury, &c., the Upper Coal-measures rest on Lower Palæozoic beds.

coal basin, the Old Red Sandstone forms bold cliffs, rising from below the Carboniferous limestone and shale, and consists of yellow sandstone and conglomerate. In North Devon, it is represented by the "Pickwell Down sandstone,"* occupying a similar position below the "Pilton and Marwood beds." In Belgium and France, it is represented by the "Psammite du Condroz," of the Upper Devonian series; and in Scotland, by red sandstone and conglomerate, unconformable to the "Lower Old Red" (or Devono-Silurian) beds. It is scarcely represented in the north of England.

(b.) *Lower Carboniferous beds.*—These immediately succeed the "Kiltorcan beds" in the south of Ireland, and there consist of grey grits and slates ("Coomhola grits") passing upwards into the "Carboniferous slate" and limestone. In the north of Ireland the "Coomhola beds," &c., are represented by massive yellow or purple grits and shales, with a conglomerate base. The Carboniferous limestone forms the greater portion of the central plain of Ireland. In Scotland the base of the Carboniferous series is called the "calciferous sandstone," and the limestone is represented by that of the Roman camp near Edinburgh. In the north of England, the "Scar limestone" forms step-like escarpments, and in Derbyshire rises into hills of 2000 feet, dipping down towards the east and west below the Yoredale beds and Millstone grit.

* Scient. Trans. Roy. Dub. Soc., vol. i. p. 147. Etheridge, Quart. Journ. Geol. Soc., vol. xxxvii. p. 196.

In South Wales the limestone forms a range of bold escarpments along the north of the great coal-basin, resting on the shales, and passing below the Millstone grit. In North Devon these shales are represented by the " Marwood," " Pilton," and " Barnstaple " beds, as already stated. The Carboniferous limestone, however, occurs in a debased form as compared with its representative further north. In Belgium the Carboniferous limestone is fully developed and underlies the coal-formation.*

(*c.*) The Table of Synonyms given on the following page may prove useful.

(*d.*) *Distribution of land and sea.*—At the commencement of the Upper Devonian stage nearly the whole of the centre and north of Ireland, the north of Scotland, the centre and north of England and Wales, were dry land; but in the southern portions of the British Isles and adjoining parts of the Continent there was an area of depression. Over the south of Ireland there appears to have been formed a freshwater lake, in which the Old Red Sandstone was deposited in the form of shingle and finer sediment drained from off the adjoining lands formed of Silurian and Devono-Silurian beds which had been previously elevated into land over the region of Kerry, Cork, and Waterford. The waters of this lake were inhabited by numerous fishes and the large mussel already named, while the adjoining lands were covered by a luxuriant vegetation, the repre-

* For a full account of the representative series given above, see Quart. Journ. Geol. Soc., vol. xxxii. pp. 613-651.

TABLE OF SYNONYMS.

	England.	Ireland.	Scotland.	Belgium.*
Upper Carboniferous	Upper Coal-measures.	(Upper absent.)	Upper Red Sandstone.	Étage Houiller.
	Middle „ „	Middle Coal-measures.	Flat Coal series.	
	Gannister beds.	Lower Coal-measures.	Slaty black-band series.	Schistes de Chokier.
Middle Carboniferous	Millstone grit.	Millstone grit or flags.	Moorstone rock.	Étage sans houille.
	Yoredale beds.	Shale series.	Upper limestone, and lower coal and ironstone series.	Grès grossier d'Ardenne.†
Lower Carboniferous	Mountain limestone.	Carboniferous limestone.	Roman Camp limestone.	Calcaire de Dinant.
	Limestone shale, or Baggy, Pilton beds.	Carboniferous slate. Coomhola grit, &c.	Upper and lower calciferous sandstone.	Schistes de la Famenne.
Old Red Sandstone (Upper).	Yellow sandstone and conglomerate, or Pickwell Down sandstone (Devonshire).	Kiltorcan beds, resting on Old Red Sandstone and Conglomerate.	Upper Old Red Sandstone.	Psammite du Condroz (lower part).

* In Belgium the lower coal-measures sometimes, but not always, rest unconformably on the limestone, the millstone grit and Yoredale beds being then absent. This was explained to me by Dr. De Koninck, at Liége.   † Dr. J. C. Purves.

sentatives of which are preserved to us in "the Kiltorcan beds." This lake may have extended eastwards into the south of England, but in France and Belgium it gave place to marine conditions, as the representative strata known as the "Psammite du Condroz" are of marine origin.* In Scotland the yellow sandstone and conglomerate, with *Holoptychius* and *Cyclopteris* (Palæopteris) *Hibernica*, was probably deposited within lacustrine waters.

On the commencement of the Lower Carboniferous stage the sea everywhere occupied the submerged tracts, bathing the sides of the uplands and mountainous parts, and bringing with it multitudes of marine animals, so that the oldest Carboniferous strata in Ireland, England, Wales, and Scotland contain numerous marine forms.† During the subsequent epoch of the Carboniferous limestone the depression proceeded, and the sea ascended on the flanks of the uplands until only the very highest elevations were left uncovered. Deep-sea conditions prevailed over the north and south of England and the centre of Ireland, and here the calcareous beds were formed in greatest thickness and purity through organic agency. A tract of country extending across England, from Shropshire, through Worcestershire and South Staffordshire, into the eastern counties appears to have remained as a ridge or land barrier, separating the

---

* Here it contains marine fossils, such as *Spirifer disjunctus, Rhynchonella pleurodon, R. pugnus*, with plants *Lepidodendron notum, Sphenopteris flaccida*, and a variety of *Palæopteris Hibernica.* Mourlon, *loc. cit.*, p. 88.

† See preceding Table of Synonyms, p. 83.

basin of the north of England from that of the south ; as the Lower Carboniferous rocks are absent, or only present as thin marginal representatives along this line of country.*

In Plate VI., Fig. 2, the relations of sea and land are indicated, as far as possible, during the middle of the epoch of the Carboniferous limestone.

It is also probable that the old rocks of the north-west of France were unsubmerged, as the little detached coal-fields of the centre of that country rest directly on these rocks without the intervention of the Lower Carboniferous beds ; at the same time, over the region lying along the borders of France and Belgium, the waters of the Lower Carboniferous sea prevailed, and the limestone formation is grandly represented.†

* The discovery of Carboniferous limestone at a depth of 890 feet below Northampton shows that the ridge was south of this spot.   Etheridge, Quart. Journ. Geol. Soc., vol. xxxvii. p. 231.

† For the fossils of the Carboniferous limestone of Belgium, see Dr. Dewalque's 'Description Géologique de la Belgique,' 2nd edit., p. 360. Recently Dr. J. C. Purves has shown that the Millstone Grit is truly represented in Belgium, in his memoir, 'Constitution de l'Etage Houiller inferieu de la Belgique' (1881).

CHAPTER VII.

(PLATE VII., FIGS. 1 AND 2.)

UPPER CARBONIFEROUS PERIOD.

THE Upper Carboniferous strata are the chief depositories of coal in the British Isles and the adjoining continental districts. They are separated from the Lower Carboniferous strata represented in Plate VI. by the middle division of the system, including the following in descending order :—*

Middle Carboniferous Series	1. The Gannister beds, or lower coal-measures.
	2. The Millstone grit, or flagstone series of Ireland.
	3. The Yoredale beds, or upper shale series of Ireland.

All the above are essentially of marine origin; those of the Upper Carboniferous series are of estuarine or lacustrine origin, with occasional marine bands at distant intervals.

(a.) *Characters of the Upper Carboniferous beds.*— The strata included under this head consist of two divisions; the lower, or " middle coal-measures," consisting of yellow and grey sandstones, blue and black

---

* This is a classification proposed in my paper " On the Upper Limits of the essentially Marine Beds of the Carboniferous Group, &c." (Quart. Journ. Geol. Soc., vol. xxxii., pp. 613–651). It has not been considered necessary to prepare a plate of this division, which would be intermediate in its arrangements between Plates VI. and VII.

clays and shales, bands of coal and ironstone. They contain plants, bivalves (*Anthracosia*, &c.), and fish remains. The occasional marine bands are to be recognised by the fossils. The "upper coal-measures" consist of reddish and purple sandstones, red and grey clays and shales, thin bands of coal, ironstone, and limestone, with *Spirorbis carbonarius*, and fish. These two divisions combined attain, in Lancashire, a thickness of 5000 to 6000 feet, but thin away rapidly in the direction of Leicestershire and Warwickshire in the Midland Counties. In Belgium these beds are also of great thickness, though the uppermost have generally been denuded away.

(*b*.) *Distribution of strata.*—The coal-measures of England and Scotland were originally distributed in two, or possibly three, large sheets, lying to the north and south of a central ridge, ranging from North Wales through Shropshire eastwards.* This I have called "the central barrier" (Fig. 2). It is uncertain whether it was not connected with the ridge of the Wicklow mountains across the Irish Channel. This old ridge may be a prolongation of a land area stretching southwards from Scandinavia, and it existed in wider dimensions during the Lower Carboniferous period.† It is also uncertain whether the

---

* The evidences of this ridge cannot here be discussed, but the reader is referred to the 'Geological Survey Memoir,' "On the Triassic and Permian Rocks of the Central Counties of England"; also to the 'Coal Fields of Great Britain,' 4th ed., p. 520. Its position was originally pointed out by Mr. R. Godwin-Austen in his important paper, "On the possible extension of the Coal-measures beneath the south-eastern parts of England," Quart. Journ. Geol. Soc., vol. xii., p. 53.

† Compare Fig. 2, in Plate VII., with that in Plate VI.

coal-measures of Scotland stretched continuously across the south of Scotland to join those of the north of England. It has been assumed that some of the higher parts of the southern uplands were uncovered by Upper Carboniferous strata, as they certainly were by those of the preceding stage. Nearly the whole of Ireland was originally covered by coal-measures.*

(*c.*) *Formation of coal-fields.*—Out of the original extensive tracts of coal-measures, almost conterminous with the boundaries of the submerged areas shown in Fig. 2, the existing coal-fields have been constructed. As compared with the original areas, their size is small indeed. This is due to the extensive denudations which took place—first, at the close of the Carboniferous period ; second, at the close of the Permian period ; and thirdly, in still more recent times. Ireland has suffered most of all, owing to the absence of Mesozoic strata ; † only small isolated patches, monuments of former more extended tracts, have been left behind.

The possible position of three coal-basins south of the Thames valley are shown in Fig. 2, very much in the position originally indicated by Mr. R. Godwin-Austen.‡   The sub-wealden boring, intended to

---

* See 'Physical Geology and Geography of Ireland,' pp. 43, 149, 163.

† *Ibid.*, p. 164.

‡ Quart. Journ. Geol. Soc., vol. xii. (1856). The same author places the line of possible coal-measures under the Thames valley, but the London borings for water do not appear to me to bear out this view. See Map No. 6, to accompanying evidence before the Royal Coal Commission.

The extension of the coal-measures beneath newer formations is indicated by the lighter shade in Fig. 1.

ascertain the nature of the Palæozoic strata along this tract, unfortunately was stopped before passing into Palæozoic rocks. The position of the coal-measures —proved under the Lias by boring at Burford—is also shown in Fig. 1, but it is impossible to determine from the data we possess the form of this coal-basin.

(d.) *Distribution of land and water.*—Little need be added to what has already been said on this point. As compared with the Lower Carboniferous epoch, the land areas became contracted owing to subsidence ; but the thickening of the strata, both towards the north-west and south-west of England, indicates the existence of extensive tracts of land, and sources of sediment, in those directions.* The waters which overspread the plains were disconnected from those of the ocean, except at intervals, though possibly at all times bordering on the sea-level of the period.

* See, on this subject, my paper " On Isodiametric Lines, &c." Quart. Journ. Geol. Soc., vol. xviii. pp. 127–146 (1862).

CHAPTER VIII.

(PLATE VIII., FIGS. 1 AND 2.)

THE PERMIAN PERIOD.

THE Permian beds are restricted to the central portions of the British Isles, and apparently were never deposited over any part of the extreme northern, western, or southern districts, or of the adjoining continental areas. According to the view of Sir A. C. Ramsay, the magnesian limestone of the north of England was formed under the waters of an inland sea, like the Baltic or Caspian, the fauna being exceedingly sparse, as compared with that of the limestones of the Carboniferous period,* and indicative of the absence of open oceanic waters. The region of the magnesian limestone of the north of England appears to have been disconnected with that of the central counties and Shropshire by a barrier ridge, the position of which is indicated in Fig. 2. To the south and west of this ridge, only Lower Permian beds are found,† and it is probable that these latter beds are, in the main, lacustrine. The Permian beds

---

* Phys. Geol. and Geog. of Great Britain, 5th ed., p. 147.

† Quart. Journ. Geol. Soc., vol. xxv. pp. 171–184; also, "Triassic and Permian Rocks, &c.," Mem. Geol. Survey, p. 10.

of Scotland are restricted to the south of that country, and those of Ireland to the districts of Down, Tyrone, and Armagh.

(*a.*) *Characters of the Permian beds.*—Owing to the dissimilarity of the Permian beds lying on either side of the Carboniferous ridge above referred to, I have arranged the Permian strata under two heads or types—those of the "Lancastrian" and "Salopian." *

The beds of the "Lancastrian type" belong to the north of England, and may thus be described in the western and eastern sections of that area :—

PERMIAN BEDS OF THE LANCASTRIAN TYPE.†

	West.	East.	
Upper Division	Bands of limestone, sometimes magnesian, with red marls. Fossils — *Turbo*, *Rissoa*, *Natica*, *Axinus*, *Schizodus*, &c.	Marls. Upper limestone. Marls. Magnesian limestone.	Fossils, marine.
Lower Division.	Lower Red Sandstone, ...	Lower Red or Yellow Sandstone.	

On the other hand, the beds of the "Salopian type" are restricted to the west and central parts of England, and consist of a thick series of red and purple sandstones, clays, or shales, with calcareous conglomerates, breccias, and boulder beds. The typical section occurs at Enville, in Salop.

* Quart. Journ. Geol. Soc., vol. xxv. pp. 171–184; also, " Triassic and Permian Rocks, &c.," Mem. Geol. Survey, p. 11.

† The exact representation of the series in Lancashire by that of Durham and Yorkshire (though somewhat dissimilar in character) proves that these beds were originally physically connected across the country, as shown in Fig. 2, Plate VIII.

The boulder beds are exceedingly like those formed by the agency of floating ice, consisting of accumulations of red, stony clay, with subangular fragments of trap, Silurian, and Cambrian rocks, some of which show surfaces slightly glaciated. Sir A. C. Ramsay considers that these breccias have been formed in waters filled with floating ice derived from lands lying towards the north-west of the submerged area.* These accumulations occur in Shropshire, Worcestershire, Staffordshire, Warwickshire, and at Armagh, in Ireland.† There are also beds of calcareous conglomerate formed of pebbles of Carboniferous limestone. The Alberbury breccia belongs to this formation.‡

(*b*.) *Distribution of land and water.*—It is probable that during the Lower Permian period two distinct basins were formed, lying on either side of the dividing ridge, both being inland lakes, or only very slightly connected with the sea. Into these lakes were carried beds of fine sand, clay, and gravel by the streams draining the adjoining lands formed of older Palæozoic rocks. On the commencement of the Upper Permian period there was a subsidence over the region of the northern lake, and the waters of the sea flowed in, bringing with them representatives of a marine fauna, and in which the great limestone beds of the north of England were deposited.

* Quart. Journ. Geol. Soc., vol. xi. p. 189.

† Phys. Geol. and Geog. of Ireland, p. 46 ; and Explan. Mem. Geol. of Armagh, Mem. Geol. Survey, sheet 47.

‡ Murchison, 'Silurian System,' p. 83.

These dolomitic limestones are unrepresented over the centre and west of England, in which the beds belong exclusively to the lower division of the Permian system, known in Germany, and now often in Britain, by the name of " Rothe todte liegende."

CHAPTER IX.

(PLATE IX., FIGS. 1 AND 2.)

THE TRIASSIC PERIOD.

THE terrestrial movements, accompanied and followed by extensive denudation, which ensued at the close both of the Carboniferous and Permian periods, produced marked changes in the distribution of the strata of the Triassic, as compared with that of the preceding periods.   There is a complete discordance between the Mesozoic and the Upper Palæozoic strata, so that the beds of the New Red Sandstone, or, in its absence, those of the Keuper Marl, rest indifferently on various members of the Permian, Carboniferous, or even older, rocks.

(a.) *Ridge of Palæozoic rocks.*—Amongst the physical changes brought about at the close of the Carboniferous period, was the formation of a ridge of Palæozoic rocks, under the south of England, of which the Mendip Hills is the western prolongation, and against which both on the north and on the south the Mesozoic strata wedge out.   Under the east of England, this ridge is in part composed of the older Silurian or Cambrian beds which occupied that district during the Carboniferous and Permian times; but it was considerably extended at the close of the Carboniferous period, and forms a portion of that great system of flexured and folded strata which range from the south of Ireland, through the south of England and Wales, into France and Belgium, and beyond the Rhine. Another ridge of great importance in physical geology is that known as the " Backbone of England," which ranged from Derbyshire northwards, and was developed at the close of the Triassic epoch. It is formed in the main of Carboniferous rocks.

(b.) *Attenuation of strata.*—It is important to observe that the Triassic strata attain their greatest development in Lancashire and Cheshire, and become attenuated in a south-easterly direction.* This is due partly to the position of the old Palæozoic ridge above referred to, and also partly to the decrease of sediment as we recede from the old lands which were the source of that sediment. From this it may be concluded that the land of the period lay to the north and west of the British Isles. It is also probable that Normandy and Brittany were portions of a land surface at the same period. The numerous beds of breccia and conglomerate in the Triassic strata of Devonshire indicate the proximity of land which may have included portions of Cornwall.† In.

* " On the South-easterly attenuation of the Lower Secondary Rocks of England." Quart. Journ. Geol. Soc., vol. xvi. p. 63 (1860).

† These beds, including the Budleigh-Salterton conglomerate, have been described by Buckland, Conybeare, and Murchison. The lower breccias were considered by these authors to be of Permian age. The section along the coast has been more recently described by Dr. Hicks, Mr. Ussher, and

Ireland, the Trias is only represented in the north-east of the country; and in Scotland, at the extreme south, and along the coast of the Moray Firth, near Elgin. It is probable that these countries were, over by far the greater part, in the position of land surfaces during the Triassic period.

(c.) *Characters of the Triassic strata.*—The Trias of Britain consist only of two divisions. The Bunter, or New Red Sandstone below, and the Keuper, or New Red Marl above; the intervening marine division of the Muschelkalk being absent in Britain.* The Bunter division consists of red sandstone and conglomerate; the Keuper of red and variegated marls and sandy shales, containing gypsum and rock-salt, with beds of sandstone and conglomerate at the base. Their basement beds originally formed in reality an old shingle beach around the flanks of the unsubmerged lands of the period.

In order to account for the absence of the middle division in Britain, I have suggested that during the formation of the Muschelkalk in Europe, the British area was converted into a land surface. The slight unconformity of the Bunter, to the Keuper, division, and the eroded surface which the former often exhibits where overlain by the latter, goes to confirm this view.

It is probable that, as suggested by Sir A. C. Ramsay, the Triassic strata of Britain were deposited

---

Mr. H. B. Woodward; the last of whom gives a good summary of the views of himself and previous authors. Geol. Eng. and Wales, p. 136 *et seq.*

* "The Triassic and Permian Rocks of the Central Counties," Mem. Geol. Survey, pp. 66 and 106.

within the margin of an inland sea or lake. The boundaries of this lake towards the north and west are inferential; those along the old ridge of the south of England have been partly determined by the aid of recent borings of the strata, which may be briefly described in the following order :—

(*d.*) *Deep borings in the centre and east of England.*— 1. Scarle, near Lincoln. The Triassic strata were reached at a depth of 141 feet, and were found to be 1359 feet in thickness.* The next (2), was at Northampton, at which the Carboniferous limestone (as determined from the fossils by Mr. Etheridge), was pierced a short distance below the bottom of the Lias, at a depth from the surface of 890 feet; † the position of this boring is evidently close to the original margin of the Trias. The next (3), was at Ware, in which the Wenlock beds were entered beneath the Gault,‡ so that the position of this boring is considerably south of the original margin of the Trias; on the other hand (4), at Burford, the Triassic strata were passed through before the coal-measures were reached, and they are also inferred by Professor Prestwich to exist under Oxford, as the water from St. Clement's well is highly impregnated with chloride of sodium. From these data the concealed line of the old ridge can be approximately drawn, while the relations of the Secondary strata along the margin of the Mendip Hills and Somersetshire coal-

* 'Coal Fields of Great Britain,' 4th ed., p. 261.

† The late Mr. Samuel Sharp has described this remarkable boring, *Northampton Herald*, 3rd Sept., 1881.

‡ Etheridge, Quart. Journ. Geol. Soc., May 1881, p. 229.

field, enable us to determine its position there with the greatest precision.*

(*e.*) *Distribution of land and water.*—The land of the period appears to have lain to the north-west, north-east, and south of the British Isles. The highlands of England, Scotland, and Ireland were certainly in the position of land, and contributed to the sediment poured into the lacustrine area. In the south the submerged area was connected with that of Normandy and Brittany; but as marginal beds rest against the eastern slope of the old rocks of Cotentin and Calvados, land probably lay over the districts of central and western France.† On the whole, the Triassic period over the region now described, was one in which elevation of land was (for that period) at its highest at the beginning, and at its lowest towards the close, when the waters of the ocean invaded the tracts covered previously by those of large lakes.‡

* This concealed ridge has been well described by Mr. Taylor in 'By-paths of Nature.' The relations of the strata of the Mendip Hills have been ably described by the late Mr. C. Moore of Bath, Quart. Journ. Geol. Soc., vol. xxiii. p. 207.

Mr. W. Jerome Harrison has recently shown grounds for believing that the pebbles of the New Red Sandstone of the midland counties of England have been derived from the old Palæozoic ridge above referred to, pp. 14 and 97, Proc. Birmingham Phil. Soc., vol. iii. p. 157.

† Mr. W. A. E. Ussher has recently visited Normandy with a view to compare the Triassic strata of that district with those of Devonshire, and arrives at the conclusion that only the Keuper division is present in the former country, the Bunter not being represented.—Quart. Journ. Geol. Soc., May 1879, p. 245.

‡ Ramsay, Phys. Geol. and Geog. of England, 5th ed., p. 155.

CHAPTER X.

(PLATE X., FIGS. 1 AND 2.)

THE JURASSIC PERIOD (INCLUDING THE RHÆTIC, LIASSIC, AND OOLITIC DIVISIONS).

AT the close of the Triassic period the waters of the ocean invaded the tracts previously covered by lakes and estuaries. The influx of these waters is shown by the fauna of the Rhætic (or Penarth) beds, which is marine, but indicates littoral and estuarine conditions. Along with *Avicula contorta*,* *Modiola minima, Pecten Valoniensis,* and *Cardium Rhæticum,* there are remains of insects, fishes, and saurians; but none of cephalopods, whose habits require an open sea. With the commencement of the Lias, however, there occurred a general subsidence of the British and adjoining European area, upon which the sea established its supremacy over all but the elevated mountainous tracts ranging from Scandinavia into Britain. The waters brought with them great shoals of cephalopods, nautili, ammonites, and cuttlefishes, as well as other inhabitants of the deep. Saurians

---

* First described by General Portlock in his 'Geology of Londonderry, &c.,' and afterwards identified by Dr. Wright, in Gloucestershire (Quart. Journ. Geol. Soc., vol. xvi.), and Mr. Bristow, at Aust-Cliff and Penarth, on the banks of the Severn.

H 2

abounded both in air, land, and water; and during the
Oolitic epoch, living forms both vertebrate and invertebrate were excessively prolific.   Although open
sea conditions generally prevailed throughout the
Jurassic period owing to subsidence proceeding more
rapidly than deposition of sediment, yet, occasionally,
shallow lagoons were formed, such as those represented
by the Stonefield and Collyweston beds; and towards
the close, those of the Portland and Purbeck beds.*

(a.) *Characters of the Jurassic strata.*—Considered
generally, the Jurassic system consists of two great
divisions.   The lower (that of the Lias), being
argillaceous; the upper (that of the Oolite), calcareous.   But this description requires modification,
as the Liassic beds contain in some places calcareous
or arenaceous strata, and the Oolite, great beds of clay
and sand.   The different divisions, as Dr. T. Wright
has shown, are characterised by different species of
ammonites, which range not only over the English
area, but into those of the Jura mountains, Switzerland, and Germany.†   The total thickness of the
group may be taken at 3000 feet, of which the
Liassic beds reach about 1000 feet.

(b.) *Distribution of land and water.*—How far the
strata of the Jurassic group originally extended, and
to what degree the higher elevations of the British
Isles were covered by the waters of the Jurassic sea
is a problem not easy of solution.   At the same time,

---

* I do not consider it necessary to enter into details and descriptions
of strata which may be found in any of the text-books of geology.

† Quart. Journ. Geol. Soc., vols. xii. and xvi.

we have several indications of the former extension both of the Liassic and Oolitic strata which go far to guide us towards some definite conclusions on this question. In the first place, having regard to the great thickness of these strata along the northern and western margins, we infer that they originally extended far beyond their present limits, and that they covered all the comparatively low-lying tracts of England now occupied by the Triassic strata. The existence of Rhætic and Liassic strata in the north-east of Ireland, and in the vale of the Eden, near Carlisle, proves the original continuity of the Liassic sea of the north with that which flowed over the central plains. These beds are also found skirting the western coast and islands of Scotland,* and, on the east, the shores of Dornoch Firth. We may well suppose, therefore, that the Jurassic sea bathed the flanks of the Irish and Scottish northern highlands, as shown in the map (Plate X., Fig. 2). The eastern limits of the tract running along the western highlands of Scotland, are in part defined by the mountains of Derry, Donegal, and the ridge of the outer Hebrides. The Atlantic area was probably distributed into ridges or islands, with intervening sea-lochs and basins, of which Rockal, Bus (or Busse), and several of the little islets or sunken rocks in that portion of the Atlantic between lat. 10° and 30° W. may be the modern indications.† The Oolitic beds of Brora,

---

* As shown by Murchison, Geikie, and Bryce. Quart. Journ. Geol. Soc., vols. xiv. and xxix.

† Geikie's Geol. Map of Scotland. Dr. W. Frazer, of Dublin, has shown a map about 200 years old, by Tassin, the Geographer Royal of France,

with coal, indicate marginal conditions along the north-eastern coast of Scotland;[*] while, as Professor Judd has shown, these beds are well represented in Sutherlandshire, and were deposited close to land; so that it is probable the Oolitic sea stretched from the north of England round the eastern coast of Scotland, during and after the period of the Lower Oolitic.

It is altogether uncertain whether the Oolitic strata were formed over the north of Ireland; but, if so, they had been swept away by denuding agencies previous to the Cretaceous period, as no strata of the Jurassic group higher than the lower beds of the Lias are found in that country.[†] It is probable that the whole of the south and west of Ireland were in the condition of land during this time.

Whether any portion of the central ridge (or "Backbone,") of the north of England remained unsubmerged during the Oolitic period is uncertain; but that the sub-Cretaceous ridge was in part (at least) uncovered by strata of this period, may be considered as highly probable, owing to the entire absence of representatives of the Jurassic period in

---

in which "the Sunken Land of Busse," now only a rock, is shown, and which was coasted by one of Frobisher's ships for three days. As Dr. Frazer has shown, the North Atlantic appears to have undergone considerable subsidence in even recent times, of which the traditional island of Hy Brasil, off the coast of Ireland, is an illustration (Journ. Roy. Geol. Soc. Irel., vol. v. n. sec., p. 128).

[*] Murchison, Tran. Geol. Soc. Lond., vol. ii., 2nd ser., p. 393.

[†] Phys. Geol. and Geog. of Ireland, p. 52.

the borings of Ware, Turnford, and London. On the other hand, the sub-Wealden boring near Battle, in Sussex, has shown that, to the south of this ridge, the Jurassic sea prevailed, and may have been deep; as the Kimmeridge clay was entered at a depth of 255 feet, and extended down to 1769 feet, below which the coralline Oolite was penetrated to a depth of 51 feet.* The sea of the south of England stretched southwards into France, and probably had its western margin in Devonshire, Cornwall, and Normandy, while its northern limits stretched from the Thames valley eastwards, to the south of Ostend, where (as already stated), the Lower Silurian rocks are found beneath the Cretaceous.

The eastern shore of the Jurassic sea was formed of old rocks which were protruded into the North Sea from Scandinavia, and stretched as far south as the Ardennes and the sub-Cretaceous ridge of the Thames valley. North of a line drawn from Frome in Somersetshire towards the mouth of the Thames, the fauna, as Professor Morris has shown, is Germanic, to the south it is Normanic.†

It will thus be seen that during the period now under consideration, the British Isles constituted a group of small islands surrounded by waters which overflowed the lower tracts and extended into the Atlantic. On the other hand, part of the existing

---

* Third Report of the Sub-Wealden Exploration Committee, by Messrs. H. Willett and W. Topley (1875), pp. 346–7.

† Quart. Journ. Geol. Soc., vol. ix. p. 317.

Atlantic Ocean itself was to some extent in the position of dry land, from which the sediment constituting the sands and clays of the Jurassic series were probably derived.*

* This view is developed in the author's paper on "The South-Easterly Attenuation of the Lower Secondary Rocks." Quart. Journ. Geol. Soc., vol. xiv. (1860).

CHAPTER XI.

(PLATE XI., FIGS. 1 AND 2.)

THE CRETACEOUS PERIOD.

(*a.*) *Terrestrial changes at the close of the Jurassic period.*—At the close of the Jurassic period, the bed of the sea was elevated into dry land over the British area, and re-distributed into lakes and estuaries, with surrounding tracts of lands formed of Jurassic and older formations. During this epoch, denudation of the strata proceeded, while beds of shale, sandstone, and limestone (representing the Purbeck formation) were deposited over the floors of the lakes. Later on, these conditions gave place to others, when the Wealden beds, restricted to the south-east of England, were deposited at the mouth of one or more rivers, draining the lands lying towards the north and west.

The interval of the Purbeck and Wealden epochs may be considered as a sort of interregnum between the great Jurassic period on the one hand, and that of the Cretaceous on the other. It was, however, one of considerable duration; and on the Continent is partly represented by the "Maestricht beds" of Belgium, and in western America, by the "Laramie beds" of Colorado.*

* According to Dr. Hayden, and Prof. Cope. Bull. U.S. Geol. Surveys, vol. v., No. 1 (1879).

Upon the commencement of the Lower Cretaceous epoch, beds of sand and gravel, now known as the "Lower Greensand" formation, were deposited in a shallow sea, and at no great distance from the land, which lay—both to the westwards (in the region now forming Cornwall, Devon, and Wales), and also along the line of the Thames valley—where the old ridge of Silurian, Devonian, and Carboniferous rocks was still uncovered by sediment.[*]

After the deposition of the Lower Greensand, there was another slight elevation of the sea-bed, and much of this formation, with portions of those below it, were swept away; but upon the commencement of the Upper Cretaceous epoch, subsidence again set in, which continued to the close of the Cretaceous period, at which time the south and centre of Europe, and all but the very highest elevations of the British Isles, were submerged beneath the waters of an ocean which must have extended eastwards from the Atlantic into Asia, and which not only occupied the basin of the Mediterranean, but the plains of France, Germany, Italy, Spain, and of northern Africa.[†] This epoch of greatest submergence is represented in Plate XI.,

---

[*] The absence of the Lower Cretaceous beds in the Crossness boring is evidence of this, as also the beds of conglomerate occurring in the Lower Greensand of Oxfordshire and Wiltshire. In the fourth edition of the 'Coal Fields of Great Britain' (p. 354), I have given a section showing the position of this ridge under London, from which it will be seen that the Palæozoic rocks were not completely covered till the period of the "Gault clay."

[†] It is probable that only a core of Palæozoic rocks of the Alps and Pyrenees were left unsubmerged at this period, but Scandinavia was probably a land area.

Fig. 2, and has been more fully dwelt on in a previous page.*

The interval of terrestrial, lacustrine, and estuarine conditions between the Jurassic and Cretaceous periods, together with the concomitant denudation, has resulted in producing a distinct unconformity of stratification between the formations themselves, along with which there is a complete change in the fauna; so much so, that with the exception, perhaps, of some Foraminifera, no species passes from the Jurassic into the Cretaceous rocks; and of 300 Lower Greensand species, only about 20 per cent. survive into the Upper Cretaceous series.†

(b.) *Characters of the Cretaceous strata.*—Speaking generally, the Lower Cretaceous strata consist of gravels, sands, and clays, of sedimentary origin, indicating a process of formation not far remote from the land of the period, and in a sea of no great depth. They are altogether absent on the borders of Devon and Dorset, where the Upper Greensand rests directly on the New Red Marl and Lias, but they commence at Devizes and extend on to Calne and Farringdon, where they are represented by the celebrated " Farringdon Gravels." On the Continent, along the axis of Artois, these beds are wanting, and the Chalk is underlain by another littoral deposit known as the Tourtia of mid-Cretaceous age.‡ The Upper Cretaceous strata indicate the prevalence of oceanic, or at least pelagic, conditions during the later stages in

---

* Page 25, *ante.*   † Ramsay, *supra cit.*, p. 217.
‡ D'Archiac, Mém. de la Soc. Géol. de France, t. ii. p. 291.

the formation of the Chalk, which is a white lime-stone composed in the main of shells of Foraminifera, and containing molluscs, crinoids, and echinoderms in great numbers. Spicules and casts of sponges are common, and are often found enclosed in flints.

The beds and nodules of flint of the Upper Chalk, are due to a process of pseudomorphism, whereby the free silica in the waters of the ocean has, from time to time, been consolidated around some body, such as an echinus, a sponge, a shell, or other foreign body, and has replaced the original carbonate of lime of which the body itself was formed, or by which it was enveloped.* The Upper Greensand formation, at the base of the Chalk, has been shown by Ehren-berg to be formed of the casts of Foraminifera pre-served in silicate of iron.† Similar casts were brought up from deep waters in the Indian Ocean by the officers of the *Challenger* expedition.

With the Upper Cretaceous beds commences the appearance of dicotyledonous plants, both in Europe and America,‡ giving a perfectly new aspect to the flora of the world, or, as it has been expressed by Dr. Oswald Heer, "introducing a new fundamental conception of the vegetable kingdom."

(c.) *Distribution of land and sea.*—At the com-mencement of the Cretaceous period, the British area

---

* The process has been fully explained by the late Dr. Bowerbank, and more recently by Professor Rupert Jones, F.R.S. Proc. Geol. Assoc., vol. iv.

† "Ueber den Grüusand," Abhand. der K. Akad. der Wissenschaft zu Berlin, 1855, p. 85.

‡ Viz. at Aix-la-Chapelle, and in America, from the Rocky Mountains to the Arctic regions at Noursoak.

was probably almost entirely in the condition of dry land, and was but slightly submerged during the formation of the Lower Greensand. Sir A. C. Ramsay considers that this submergence was so slight that the Oolitic strata, which then extended far to the west into the borders of Wales, were not entirely submerged.* After its deposition, the land was tranquilly raised out of the sea, and subjected (along with the older strata) to atmospheric waste.

The deposition of the Gault in England first took place while the surface of the country was being gradually submerged, and part of the sediment was distributed over the Lower Greensand, or along the flanks of the little ranges of hills formed out of it, and part over the underlying Oolitic strata. As time went on the submergence increased, and more rapidly than the filling up of the sea-bed by the accumulation of Upper Cretaceous strata. During the period of the Upper Chalk, the submergence reached its maximum. I have already stated the great extent of the existing land surface over which the ocean waters spread in the centre and south of Europe. The submergence of the north of Ireland, and of, at least, the borders of the Scottish highlands is indicated by the presence of the Chalk and Upper Greensand overlying the Lias in County Antrim, and by similar beds in the Isle of Mull, and at Bogingarry, in Aberdeenshire.† To what extent

---

* *Supra cit.*, p. 230.

† As shown by Prof. Judd (Quart. Journ. Geol. Soc., vols. xxix. and xxx). Professor Judd considers that the Cretaceous beds once " extended over large portions of Scotland," from the presence of chalk flints beneath

this submergence progressed is of course uncertain, but we may assume that the more elevated districts formed of Palæozoic rocks were not completely under water, while it is highly probable that land lay over a large tract of the Atlantic, extending westwards from the Scandinavian promontory, as I have endeavoured to represent in Fig. 2, Plate XI. The highlands of Cumberland, Wales, and Ireland were also, in all probability, in the condition of land surfaces. It is more uncertain what was the condition of such tracts as those of Dartmoor, in Devonshire, the southern uplands of Scotland and the Isle of Man.*

The relations of land and sea shown in Plate XL, Fig. 2, are those which are supposed to have existed during the formation of the Upper Chalk. The more deeply submerged areas, extending into France and Belgium, are shown by a deeper tint of blue—the more elevated unsubmerged mountain tops by correspondingly deeper tints of brown.

---

the basalts of Mull, &c., as well as from the occurrence of detached outliers both on the mainland and in several of the western islands, *ibid.*, vol. xxix. p. 105.

* The distribution of land and sea, as shown in Fig. 2, very nearly agrees with the ideas stated by Professor Sir A. C. Ramsay in his ' Physical Geography of Great Britain,' 4th ed., p. 257.

CHAPTER XII.

(PLATE XII., FIGS. 1 AND 2.)

THE TERTIARY PERIOD (EOCENE, OLIGOCENE, AND MIOCENE DIVISIONS).

(*a.*) *Distribution.*—The Tertiary strata of the British Isles are restricted to the southern parts of England, the north-west of Scotland, and the north-east of Ireland. I have represented on the map (Fig. 1, Plate XII.) the position of the deposits belonging to the Eocene, Oligocene, and Miocene divisions. In dealing with the physiography of these deposits we will consider the Eocene and Oligocene in the first instance, and the Miocene in the second.

(*b.*) *Eocene and Oligocene Strata.*—These deposits occur chiefly in two separate tracts or " basins "—that of London, and that of Hampshire and the Isle of Wight. They also occupy a large tract of the adjoining continent. Originally these were connected in one great sheet ranging into the centre of France, and extending in England far beyond their present limits, both northwards and westwards, but *how far* it is extremely difficult, if not impossible, now to determine. Their disseverance into three separate tracts took place during the Miocene period; when

—by the contraction of the earth's crust, the elevation of the ground now occupied by the Weald of Sussex into an anticlinal arch bordered by corresponding depressions, and the subsequent denudation of the strata—the London Tertiary basin was separated from that of Hampshire, and the Cretaceous strata, with the underlying Wealden beds, were brought to light. At the same period, by denudation along the Straits of Dover, the Paris Tertiary basin was dissevered from that of London. This process of denudation has proceeded ever since, and has affected not only the Tertiary but the subordinate Secondary strata of the Cretaceous and Jurassic series: so that vast tracts of chalk, greensand, oolite, and lias have been denuded away by atmospheric agencies from the commencement of the Miocene epoch downwards, and the boundary scarps have receded farther and farther in the direction of the dip of the strata to their present positions over the whole of the south of England, and the adjoining district of France and Belgium.

During the formation of the Eocene and Oligocene beds, the north and west of the British Islands was, in all probability, in the condition of dry land. It is probable that Ireland was joined to England and Scotland by a tract of Cretaceous rocks mantling round the hills of older formation. To the south of this tract of land the waters of the Tertiary sea were spread, extending over the north and centre of France and Belgium, in which direction they became more limpid and free from sediment than during the epoch represented by the London clay; so that while beds of clay were being deposited over the area of the estuary of the Thames, others of pure limestone with *Nummulites* * were being formed over the area of the Paris basin.

(*c.*) *Characters of the Eocene and Oligocene strata.*— The oldest beds consist of gravels, sands, and clays, of marine or fluvio-marine origin.† These are suc· ceeded by the London clay, a blue and brown stiff clay, with *Septaria*, about 500 feet in thickness, and of marine origin. This formation thins away in the direction of the Isle of Wight. The succeeding Middle Eocene beds consist of the Bagshot sands and Bracklesham beds, of estuarine origin; and the upper, of the Barton clay, of marine origin, 300 feet in thickness. The "Oligocene beds" of Professor Beyrich have been shown by Professor Judd to be present in the Isle of Wight,‡ and consist of alter- nating sands, clays, shales, and limestones of marine and estuarine origin.

All these beds were deposited either at the mouths of rivers flowing from the north and west into a sea, which was generally open in the direction of France, but often very shallow, and sometimes converted into estuaries, and even lakes, of limited extent. Oceanic

* In the lower part of the "Calcaire grossier" of Paris there are three species of *Nummulites* abundant, viz.: *N. lævigata, N. scabra, N. Lamarki,* which serve to show that it corresponds to the epoch of the formation of the great Nummulite limestone of the south of Europe, &c. One of these, *N. lævigata,* occurs in the Middle Eocene beds of England.

† The Thanet sands and "Woolwich and Reading beds" of Prestwich.

‡ Quart. Journ. Geol. Soc., 1880.

water, such as that of the Atlantic, probably never occupied the area in question. The fauna indicates successive stages of depression or elevation, and the alternation of fresh - water, estuarine, or marine conditions.

(*d.*) *Distribution of land and sea.*—I have already to some extent dealt with this subject, and will, therefore, only here observe that, as a long interval of time elapsed between the formation of .the Upper Chalk and of the Lower Eocene strata,* during which land conditions prevailed over the British area, much of the Chalk formation itself was denuded away; and consequently the Tertiary beds are unconformable to the Cretaceous, and rest sometimes on higher, sometimes on lower strata of that formation. Of a similar character, also, are the relations of these formations to one another over the Franco-Belgian area, where the Nummulite beds are found to rest discordantly on those of Cretaceous age.†

The position of the northern range of the sea margin, even during any given epoch of the Tertiary period, is a question of much uncertainty. In Fig. 2, Plate XII., I have attempted to show the physical geography of the epoch of the London clay, when the sea had its greatest extension over the area here described. At the same time, in the absence of special indications for guidance, I have considered it necessary to leave a considerable tract of uncoloured

---

* This interval is partly represented by the Maestricht beds of Belgium.

† Godwin-Austen, *supra cit.*, p. 72.

debatable ground between the respective margins of land and sea.

(*e.*) *Miocene strata.*—With the exception of some lacustrine beds of gravel, clay, and lignite at Bovey Tracey in Devonshire, all the British representatives of the Miocene epoch are restricted to the north-east of Ireland, and the west coast and isles of Scotland, where they are of volcanic and lacustrine origin.

These beds consist of great sheets of augitic and felspathic lavas, with intervening beds of ashes, lapilli, pisolitic iron ore, and lignite with plants, the examination of which enabled the late Professor Edward Forbes to determine the Miocene age of these rocks. These volcanic sheets rest generally on a floor of Chalk, or of some older formation, and observers are agreed that they have been poured out upon a land surface. It is probable that, at one or more intervals, lakes were formed, and that the valuable pisolitic iron ore of Antrim may be referred to a lacustrine mode of accumulation.

(*f.*) *Denudation.*—These volcanic products have undergone enormous denudation since the Miocene period, and on the little map, Fig. 2, Plate XII., I have endeavoured to show the original area overspread by them. If this be compared with their present superficial area the extent of this denudation will be understood. Much of this destruction of the volcanic products took place before the Pliocene period, for we find the Pliocene clays of Lough Neagh resting against a shelving bank of the basaltic sheets.* Even previous to the glacial period many of the principal valleys, glens, and escarpments, had assumed their main outlines although hollowed out of hard sheets of lava, and much of the newer formations had been bodily removed.

* Phys. Geol. of Ireland, p. 72; also, Explanatory Memoir to Sheet 35 of the Geol. Survey of Ireland, p. 13, &c., by E. T. Hardman.

CHAPTER XIII.

(PLATE XIII.)

1. THE GLACIAL, OR POST-PLIOCENE PERIOD.

(*a.*) *Three main divisions.*—The Glacial or post-Pliocene period has been generally, and, as I believe, correctly, distributed into three distinct epochs, which merged into each other, but were each of prolonged duration. During each epoch the climatic conditions, the relations of land and sea, and the resulting deposits, were different, and may be briefly tabulated as follows for the area of the British Isles.

THE GLACIAL PERIOD.

Epoch or Stage.	Terrestrial Conditions.	Climatic Conditions.	Formations.
3. Upper.	Partial submergence.	Sub-arctic.	Upper Boulder Clay.
2. Middle.	Deepest submergence.	Temperate.	{ Middle Sand and Gravel.
1. Lower.	{ Greatest elevation of land.	} Arctic.	{ Lower Boulder Clay or Till.

Between these deposits and the Norwich crag are some interesting glacial or sub-glacial beds, indicating the approach of the arctic conditions which prevailed during the formation of the Lower Boulder Clay.*

* These deposits are included in Mr. S. V. Wood's "Lower" and "Middle" Glacial series; but as Dr. J. Geikie has shown, they are of older date than the three stages given above.—'Great Ice Age,' p. 370 (1874).

(*b.*) *The British Area during the epoch of the Lower Boulder Clay.*—Plate XIII. represents the physical conditions of the British area during the Lower Glacial stage, represented by the Lower Boulder Clay (epoch or stage 1), and the general glaciation of the exposed rock surfaces. It will be observed that the whole of the present area of the German Ocean, as far south as lat. 51° 30', was filled with a great ice-sheet, stretching southwards from the Scandinavian peninsula,* which at that epoch was covered, like Greenland at the present day, with a continuous sheet of snow and ice. This ice-sheet became divided into two divergent sheets in lat. 57° 30' owing to the obstruction to its course caused by the position of the Scottish highlands, and the large masses of ice descending in an easterly direction from the snow-fields of the Grampians. While one portion took a south-westerly course towards the Norfolk coast, another moved in a direction perpendicular to this, and passing over the Orkneys † and the northern end of Caithness in a north-westerly direction, ‡ protruded outwards into the Atlantic. § The Scandinavian ice-sheet, however, does not appear to have extended to the Faröe Islands, which, as Dr. J. Geikie has recently shown, were glaciated by ice

---

* Croll, ' Climate and Time,' p. 444.

† Peach and Horne, Quart. Journ. Geol. Soc., vol. xxxvi. p. 648.

‡ T. F. Jamieson, *ibid.*, vol. xxii. p. 261.

§ Dr. Croll, *ibid.*, map, p. 449. This author takes the margin of the ice-sheet much farther westward in the Atlantic than that shown in Plate XIII. It would be impossible to determine the true limits, which in all cases must be hypothetical.

having a strictly local origin amongst the central heights of these islands themselves.* The great Scandinavian ice-sheet, however, was joined by another descending from the snow-fields of the northern highlands, which passed right across the Minch and over the lower parts of the outer Hebrides into the Atlantic. Large masses of ice descended in a southerly and westerly direction from the mountains of Perthshire and Argyllshire, as indicated by the glacial striæ of the rock surfaces, and uniting with that of the southern uplands of Scotland,† passed westwards across Cantyre and the North Channel. From the western and northern coast of Ireland the ice likewise protruded seaward, so as to form with that of Scotland a nearly continuous sheet, as indicated by the rock-striations.‡ The whole of Ireland was covered by an ice-sheet, moving from an axis which stretched from the neighbourhood of Lough Corrib in the south-west, to Lough Neagh in the north-east.§ This mass was augmented by others of smaller size and extent, descending from the local snow-fields of Donegal, Galway and Mayo, Cork and Kerry, Waterford and Wicklow. The whole of the Irish Sea, as far south as lat. 52°, was probably filled with ice, coming from Ireland on the one hand, and from the south of Scotland and the north of England on the other. The ice moved across Anglesea in a

---

* Trans. Roy. Soc. Edin., vol. xxx. p. 217.
† J. Geikie, 'Great Ice Age,' plate xv. and text.
‡ Rev. M. Close, Journ. Roy. Geol. Soc. Ireland, vol. i.
§ Phys. Geol. and Geog. of Ireland, p. 225, and map, p. 211.

S.S.W. direction,* and along and over parts of the Isle
of Man in a nearly parallel course.† The course of
the ice-path in Lancashire and Cheshire is indicated
on the map by the arrows.‡ The glaciers of North
Wales were not of sufficient magnitude to deflect the
northern ice from its course, but only augmented its
volume. Along the banks of the Mersey, at Liver-
pool, the direction of the ice flow was S. 35°, E.§
The exact southern limit of the ice-sheet across
England is uncertain, but it is probable that it
ranged across somewhere south of Welshpool, Shrewsbury, and Birmingham. In the centre of England the
northern ice-sheet came in contact with that from
Scandinavia, the presence of the latter being indicated by the deposit known as the "chalky boulder
clay." ‖ It is probable the high district of the
Carboniferous Limestone and Millstone Grit of
Derbyshire and Lancashire was not overflowed by
the ice; the same observation applies, with less of
certainty, to the ridge of the Cleveland Hills, and of
the Chalk, north of the Wash; west of this ridge the
"red boulder clay" of the northern ice-sheet is distributed. The direction of the ice movement along
the N.E. of England has been noted by Sir A. C.
Ramsay,¶ and that along the valley of the Forth, by

* Sir A. C. Ramsay, Phys. Geog. of Great Britain, 4th ed., p. 403.

† Rev. G. J. Cumming, Quart. Journ. Geol. Soc., vol. ii. Mr. Cumming
refers the glacial phenomena to the action of floating ice.

‡ R. H. Tiddeman, Quart. Journ. Geol. Soc., vol. xxvii. p. 490; and
J. G. Goodchild, *ibid.*, vol. xxxi. p. 55.

§ G. H. Morton, Rep. Brit. Assoc., 1870.

‖ S. V. Wood, Quart. Journ. Geol. Soc., vols. xxiii. and xxxvi.

¶ *Supra cit.*

Professor Geikie and his brother.* The sheets of ice
descending from the Grampians were met by those
descending from the snow - fields of the southern
uplands, and both united took an eastward course till,
being opposed by the heavy masses of Scandinavian
ice blocking the North Sea, the stream was deflected
southwards and passed along the north-east coast of
England. The extreme southerly margin of the ice
was limited by its melting, and doubtless numerous
muddy streams issued forth at its base, while the
Atlantic was filled by large bergs breaking off at
the ice-foot and floating southwards with the oceanic
current, as in the case of the Greenland sea at the
present day.

Mr. W. Keeping has recently given valuable
information regarding the Glacial deposits of central
Wales.†

In the south-east of England, the Lower Boulder
Clay of Lancashire was preceded by beds of gravel
and clay with erratics, constituting Mr. S. V. Wood's
" Middle " and " Lower Glacial " series, and by
which the great Chalky Boulder Clay, the. true
representative of the Lower Boulder Clay of
Lancashire, is separated from the Norwich Crag.‡

(c.) *Elevation of land.*—There is reason to believe
that during this early stage of the Glacial period

* Geological Map of Scotland, 'The Great Ice Age,' &c.
† Geol. Mag., No. 216 (June 1882).
‡ I agree with Dr. J. Geikie in considering Mr. Wood's Cromer series
as preceding the Lower Boulder Clay of the west of England, and as
representing probably the true Till, with fresh-water beds lying at the
base of the Glacial series of Scotland.

the land was elevated, and that much of the shallower portions of the sea-bed were laid dry.   Under these circumstances the south of England would have been united to France (as shown in Plate **XIII.**), while the North Sea would have been shallow.   But, as Dr. Croll believes, the mass of ice from Scandinavia was so great that it took possession of the North Sea, dislodging the waters which were insufficient in depth to break it up, and float it away in the form of bergs.

(d.) *Directions of ice flow.*—As regards our knowledge of the direction of the ice flow, the evidence is mainly of two kinds, that derived from the lines and groovings found on rock surfaces *in situ*, and that derived from the nature of the Boulder Clay, or Till, and the stones or boulders it contains.   It being assumed that this deposit has been formed by the ice-sheet, the stones which it contains can often be traced to their sources, and thus the direction of the ice movement becomes known.*

On the other hand, erratic blocks strewn over the surface are not to be relied upon as evidence of the former presence of an ice-sheet, as in many cases they have been carried by floating ice from their original sources at a time when the country was partially submerged.   The periods of submergence followed that of the great ice-sheet, and are illustrated in the succeeding maps, in Plate **XIV.**, Figs. 1 and 2.

---

* Mr. D. Mackintosh has ably carried out observations of this kind, which are to be found in communications to the Geological Society of London.

(PLATE XIV., FIG. 1.)

2. INTER-GLACIAL EPOCH.

(*a.*) The inter-glacial stage of the post-Pliocene period presents in the main a remarkable contrast to that which preceded it. Instead of intense cold, there was a temperate climate similar to our own; instead of elevation of the land, there was deep depression and extensive submersion beneath the waters of the sea; and instead of the formation of a glacial deposit like the boulder clay, there was the deposition of beds of sand, gravel, and loam, often containing sea-shells identical with existing species. The physical conditions of the two epochs could scarcely have been more different over the area here described, but this difference was by no means confined to the limited region of the British Isles. It extended, as Dr. Oswald Heer has shown, to Switzerland and the centre of Europe, and as Dr. Dawson has shown, to North America.

The occurrence of an inter-glacial stage between two others of a glacial character, is admitted by most writers on the physical history of post-Tertiary times.* The representative beds, or formations, have been recognised both in the west and east of England; but some observers, including Dr. J. Geikie and Mr. S. V. Wood, consider these three-

* Dr. James Geikie, 'Great Ice Age,' 2nd ed., p. 328; Sir C. Lyell, 'Antiquity of Man,' 4th ed., p. 259–60. Dr. Geikie and Dr. Croll consider there were several inter-glacial stages; but, however this may have been in Scotland, only one can be recognised in England and Ireland with certainty. The more mountainous character and higher latitude of Scotland may account for many of the peculiarities of its glacial history. The change from one set of conditions to another was doubtless gradual and slow.

fold stages not to be representative of each other in time, but to some extent post-Tertiary.   This is a view which, upon careful consideration of the subject, I am satisfied is based upon good grounds. In the east of England, the post-Tertiary deposits in succession to the Norwich Crag are so clearly and fully represented, that they enable us to trace the successive stages of the earlier glacial epoch in a manner of which we have nowhere else a parallel. Agreeing with **Dr. J. Geikie**, that there are beds in Norfolk, intermediate between those of the Glacial epoch and the Crag, not represented in the west of England, the following appears to be the succession of these deposits, with their western equivalents, as made out by Messrs. S. V. Wood and J. L. Rome :—

(b.) *Representative Post-Pliocene Series.*

Lancashire, Cheshire, &c.	East Anglia.
Upper Boulder Clay	Hessle clay with boulders.
Middle Sands and Gravels (marine) .. .. {	Marine gravels, giving place to littoral and fresh-water gravels.
Lower Boulder Clay .. .. {	Purple Boulder Clay of Yorkshire. / Ditto with chalk. / Great Chalky Boulder Clay.
Earliest Glacial beds, represented in the north of England and Scotland by the "True Till," with fresh-water beds	Sand and rolled gravel with shells.* / Contorted drift with masses of marl and chalk. / Boulder clay with erratics.† / Laminated blue clay. / Fluvio-marine sand and clay.
Pre-Glacial Beds..	2. "Forest bed of Cromer." / 1. Sand and gravel with loam (Norwich Crag). / Chalk.

---

* Mr. Wood's "Middle Glacial beds."
† Mr. Wood's "Lower Glacial beds."

Taking a general view of the subject, it must be supposed that the glacial and sub-glacial beds overlying the forest bed of Cromer were the precursors of the great land ice-sheet, represented by the Chalky Boulder Clay, and that the marine gravels overlying the Purple Boulder Clay are the representatives of the middle sand and gravels of Cheshire, Lancashire, Wales, Wicklow, and Wexford.*

(c.) *Distribution of inter-glacial beds.*—The beds formed during the inter-glacial stage above described are widely distributed, and consist of sand, gravel of water-worn pebbles, and beds of loam, generally finely laminated; they give evidence of deposition under water. The occurrence of sea shells in various parts of the British Isles proves that they lived in the waters of the sea, and the genera and species vary somewhat according as warm or cold currents were present. Sometimes the gravels are found resting directly on the Lower Boulder Clay, as at Howth and Killiney in Ireland, on the banks of the Ribble near Preston, and in East Anglia. In other places they rest directly on the solid floor of the older rocks. They also cover large tracts of the central plain of Ireland, ascend the Wicklow mountains to an elevation of 1235 feet, as shown by the Rev. Maxwell Close, and occur along the eastern coast of Ireland, often over-

---

* The presence of *Fusus contrarius*, and two or three other forms supposed to have been extinct, but represented in the Suffolk Crag, induced the late Professor Edward Forbes to refer the Wexford gravels to the age of the Crag, but this shell has recently been dredged up off the coast of Spain as a "living" form. Professor Haddon has shown me a specimen from the collection of the Royal College of Science, Dublin.

laid by the Upper Boulder Clay. Amongst the Sperrin mountains in Ulster they have been found by Mr. Joseph Nolan at an elevation of 1200 feet. Amongst the mountains of North Wales . they have been detected by the late Mr. J. Trimmer,* by Mr. D. Mackintosh and others at somewhat higher elevations, and by Professor Prestwich on the hills near Macclesfield, at an elevation of about 1100 to 1200 feet.† They are spread over large tracts of Lancashire, Cheshire, and Salop, sometimes occurring at the surface, but as often concealed beneath the Upper Boulder Clay. In the central counties they are extensively distributed, and on the tableland of the Cotteswold Hills of Gloucestershire and Oxfordshire, I have traced them to elevations of about 600 feet. ‡ The "high-level" gravels of Berks, Wilts, Dorset, and Oxfordshire are also probably referable to this division of the drift series. Representative beds are present in the east of England, interposed between two boulder clays.§ In the Lake District of Cumberland the late Rev. C. Ward has traced stratified gravels up to at least 1500 to 1600 feet.‖ In the west of Scotland I have found similar beds of gravel and sand high up amongst the hills of

* On Moel Tryfaen at 1360 feet. The shells have subsequently been named by Mr. R. D. Darbyshire.

† Lyell, *in loc. cit.*, p. 317.

‡ Quart. Journ. Geol. Soc., vol. xi. p. 477. Dr. J. Geikie places the submergence in Scotland at not less than 526 feet, or much more (*ibid.*, pp. 163 and 329). Lyell suggests 2000 feet for Scotland.

§ See p. 124.

‖ "The Geology of the Northern part of the Lake District," Mem. Geol. Survey, p. 94.

Cantyre, and at less elevations in the neighbourhood of Glasgow. Out of these gravel beds the more recent Eskers or Kames appear to have been constructed. The south of England was probably only very slightly submerged at this stage.

(d.) *Amount of submergence.*—Such then is a general account of the distribution of these interglacial beds. The elevation at which the gravels are found is assumed as an index to the measure of submergence of the land, as they were certainly formed *in situ* amongst the mountains. This submergence probably reached its maximum of 1300 or 1400 feet about the centre of the British Isles, along the parallel of 53° N. lat., and was less in the south of England, and perhaps in the north of Scotland. At the time the shelly gravels were being deposited the British Isles became an archipelago in miniature. In the little map, Plate XIV., Fig. 1, I have endeavoured to represent their condition during the epoch of greatest submergence.

* This may be compared with that of Lyell (*in loc. cit.*, p. 325).

(PLATE XIV., FIG. 2.)

3. EPOCH OF THE UPPER BOULDER CLAY.

(*a.*) *Changes of physical conditions.*—The epoch of greatest submergence, represented in Plate XIV., Fig. 1, when marine gravels were deposited on mountain slopes of the British Isles at elevations as high as 1360 feet, and when glacial conditions disappeared, except perhaps amongst the islets formed of the summits of the Scottish highlands, was succeeded by a second epoch of glacial conditions; not, however, as severe as the first, and one which took place when the lands were to a less extent submerged. After the pause accompanying the deep submergence above referred to, the land began to rise, and considerable tracts of mountainous and hilly ground, previously overflowed by the sea, reappeared, and were converted into dry land. This uprising was accompanied by a return of cold, so that *small* snow fields giving birth to *smaller* glaciers began to accumulate on the higher elevations; and, as the glaciers in some cases entered the sea, small bergs and rafts of ice dotted the surface of the water, and carried their freights of boulders, stones, clay and sand in the direction towards which they were impelled by the winds and currents of the period, and as they melted dropped their loads over the submerged tracts. At the same time, owing to the melting of the snow or ice, numerous streams of red, muddy, glacier water entered the sea, which must have been thus discoloured over the central and northern portions of the area referred to. From this red mud a deposit would be formed, we may infer, similar to that of the Upper Boulder Clay of Lancashire, Cheshire, North Wales and Ireland, the Hessle clay of Yorkshire,* and its representatives in the north-east of England.

(*b.*) *Character of the Upper Boulder Clay.*—This deposit consists of reddish clay, slightly laminated, and containing bands of sand or loam. In some places it contains foreign stones and small boulders brought from a distance, and in a few instances marine shells have been detected in it. Thus at Gorton, near Manchester, the following rock fragments, nearly all foreign to the neighbourhood, were determined some years ago;—†

	Per cent.
Silurian grit	37
Felspar porphyry	31
Felstone	2
Carboniferous grit	14
Granite	6
Porphyritic agglomerate	4
Carboniferous limestone	3
Ironstone	2
	99

The majority of the above specimens had evidently been transported from the district of Cumberland

---

* This is described by Mr. S. V. Wood as a deposit of clay containing a few scattered stones and boulders formed when the sea extended over the land to an elevation not exceeding 350 or 400 feet anywhere in Yorkshire (Geol. Magazine, vol. vii.).

† By Sir A. C. Ramsay and the author (Geol. of Oldham, &c., Mem. Geol. Survey, 1864).

K

and North Lancashire, which we may suppose sent off loads of stones and boulders southwards upon icebergs and rafts.

(c.) *Localities.*—The Upper Boulder Clay rises to elevations of 500–600 feet amongst the western slopes of the Lancashire hills, and marine shells (*Turritella terebra, Fusus Bamfius, Purpura lapillus,* &c.) have been found in it; as, for instance, at Hollingworth Reservoir (568 feet above the sea), the vale of Mottram, Bradbury and Hyde.* From this elevation it gently slopes southward and westward, towards the plain of Cheshire,† which is largely overspread by it; it occupies portions of the low valleys of North Wales.‡ On the other hand, the Pennine table-land of South Yorkshire and North Derbyshire, and the low country to the east of it are free from drift deposits,§ a state of things very difficult to explain, but clearly indicating the absence of glacial conditions amongst the eastern valleys of the Pennine chain south of the parallel of 53° 35′ N. lat.

(d.) *North and East of England.*—The Upper Boulder Clay referred to above, indicates the recurrence of glacial conditions, but not to the extent of those which prevailed during the formation of the Lower Boulder Clay or Till. In Lancashire · and

* By Mr. Bateman, C.E., Professor Prestwich, and Mr. John Taylor (Geol. of Oldham, &c., Mem. Geol. Survey, p. 51, 1864).

† Geol. of North Derbyshire (*ibid.*, p. 75).

‡ In a pit by the railway side, near Abergele, it may be observed capping the inter-glacial gravels at an elevation of only 20–30 feet above high water.

§ Geol. of Dewsbury, &c., Mem. Geol. Survey, p. 20 (1871).

Cheshire this deposit may be seen resting on the inter-glacial gravels and sands in the banks of the Ribble above Preston, and on the coast near South-port, as well as in many other places. It consists of red clay with stones and small boulders often glaciated, but the clay is laminated, and was evidently formed under water. In the east of England, however, it seems to be represented by a second formation of boulder clay known as "the Hessle Clay with boulders," described by Messrs. Wood and Rome.*

As regards the Lake District of England, this epoch appears to be referred to by the late Mr. Ward in the following passage:—"The submergence [which occurred during the formation of the sands and gravels] continued until the land had sunk more than 2000 feet below its present level, as the position of boulders in many parts of the district seems to show, and notably those on Starling Dodd. Then the whole district was represented merely by scattered islands clad in snow and ice, each a little nursery of icebergs."†

(e.) *Ireland.*—The Upper Boulder Clay, resting on the marine gravels of the inter-glacial stage, has been noticed in several places, as at Killiney near Dublin, along the Wexford coast,‡ at the marble quarries near Kilkenny,§ at Modubeagh colliery near Car-

---

* Quart. Journ. Geol. Soc., vol. xxiv. Mr. S. V. Wood has contributed still more recently another elaborate account of the glacial beds of this part of England.   † *Supra cit.*, p. 94.

‡ Professor Harkness, Geol. Magazine, vol. vi. p. 542.

§ Phys. Geol. Ireland, p. 90. Geikie, 'Great Ice Age,' 2nd ed.

low,* and in counties Tyrone, Antrim, and Derry. It is similar to its English representative, but has probably suffered more from denudation, so that it is only to be found in small detached areas. When it was in course of formation the land was probably depressed to a level of about 1000 feet below that it now occupies, and as the sea-bed still farther rose, the soft material of which it was composed would have offered but slight resistance to the waves and currents which chafed around the unprotected prominences. In more than one instance which has come under my notice the formation would seem only to be represented by blocks of travelled stone stranded on the surface. An instance of this kind occurs at Kilkelly, in co. Mayo, where large slabs of Carboniferous grit are to be found strewn over a tract of country of considerable extent, covered by a thick deposit of gravel on which these blocks are found resting.†

(f.) *Scotland.*—In this case we are met by difficulties of identification, as the geologists of that country do not seem to have recognised a representative to the Upper Boulder Clay of England.‡ Dr. J. Geikie refers to the "Upper Drift Deposits," very diverse materials, such as coarse, earthy débris

---

* Hardman, Journ. Roy. Geol. Soc., Ireland, vol. iv. p. 73 ; Expl. Mem., Sheet 35 of the Maps Geol. Survey. Here the Upper Boulder Clay was proved to be 84 feet in thickness, resting on 25 feet of sands, gravels, and clays, and this again on Lower Boulder Clay 8 feet. The elevation is about 750 feet.

† Originally described by Sir R. Griffith, Brit. Assoc. Rep. 1844.

‡ My own observations in the Glasgow district, however (1868–9), lead me to think that such a deposit may occur east of that city.

of angular fragments, and large blocks and boulders strewn over the northern slopes of the southern uplands. These he traces to the Grampians as their source, and considers they have been brought to their present position on a second great sheet of ice moving southward.* It is, of course, possible that the more northerly position, the greater elevation of the Grampians and north highland mountains than those of other British districts, and the consequently greater amount of snow and ice which must have accumulated on their summits and slopes, may have produced a second ice-sheet which has no representative elsewhere; and in such a case, the central valley, though really below the sea level to an extent perhaps of several hundreds of feet, may have been completely filled with ice, which for a time excluded the waters of the sea.† But admitting all this, it is inevitable that when the ice began to give way, owing to the approaching amelioration of the climate, it would be broken up into rafts and bergs answering in all respects the description given of these phenomena as they are supposed to have occurred in England and Ireland.

(g.) *Local moraines of this epoch.*—The local

---

* *Supra cit.*, p. 120.

† Dr. Geikie, and also Mr. Jamieson (Quart. Journ. Geol. Soc., 1865) have treated very fully of the formation of kames (in Ireland called "eskers"), which the former refers to the upper glacial deposits, but I prefer, for reasons I have stated elsewhere (Phys. Geol. of Ireland, p. 100), to regard them as post-glacial. The phenomena in Ireland are very similar to those of Scotland.

moraines which existed amongst the higher unsubmerged districts, became centres of dispersion of
erratics which were floated to their destination on
masses of glacier ice. If (as Dr. J. Geikie considers)
about the time now referred to, the south of Scotland
was submerged to the depth of 1100, or even 1250
feet,* and the north of England and of Ireland to a
depth of about 900 or 1000 feet, the tract submerged
would be very large, and boulders would be carried
in directions corresponding to the prevalent winds
and currents. The courses travelled by such erratics
have been indicated by, amongst others, Mr. D.
Mackintosh, who has traced the boulders over large
tracts of country in the north-west and centre of England and Wales to their parent masses.† Amongst
the more remarkable instances of erratic blocks are
those at Pagham and Selsea, carried hither from
Normandy, probably by floating ice.‡

(h.) *Gradual disappearance of glacial conditions.*—
It is unnecessary for my present purpose to go more
fully into the details of later glacial and post-glacial
phenomena. It will probably suffice that I should
add that the climatic and geographical conditions of
the stage of the Upper Boulder Clay gradually gave
place to those which preceded, and ultimately intro-
duced, the existing temperate conditions of climate.
The land gradually rose out of the sea, the rise being
probably accompanied by prolonged pauses. The

* From Mr. H. M. Skae's observations in Nithsdale.
† Quart. Journ. Geol. Soc. London, August 8, 1879, p. 425.
‡ Lyell, 'Antiquity of Man,' p. 280.

snows and glaciers melted off the mountains. The sea was gradually freed from ice, and the waters became pure and limpid. The plants and animals of the adjoining continent once again flocked over and restored life and verdure to the face of nature. Man himself followed in their train, and made his dwelling in the caves of the rocks, living by the chase, and trying his strength with some of the fierce carnivores which infested the forests and dens of the mountains.* With this state of affairs geology closes its record, and makes way for the researches of the antiquarian and historian.

It will be observed that in the maps referring to the glacial period (Plates XIII. and XIV.) I have represented only the supposed physical restorations of the surface of the country as they were during the three special stages to which they point. I have not attempted to produce corresponding maps showing the distribution of the various glacial deposits. To attempt this would have been impossible on a scale so small as those of these three little maps, even if I had had the necessary materials to guide me. But such is not the case. The mapping of the Quaternary deposits in detail has as yet been only partially carried out by the Government surveyors, or by private agency, and it is of such a character that it

---

* I place the advent of man as post-glacial deliberately, as Dr. John Evans, F.R.S., our highest authority on such questions, has recently analysed the evidences which have been adduced both in Europe and the British Isles, for assigning to him a pre-glacial advent, and finds them in all cases more or less untrustworthy.

could not be effectively reproduced on a single map unless one on a very large scale.*

GENERAL CONCLUSIONS.—We have now passed in review the successive geological stages as represented by special epochs of which I have attempted to restore the physical outlines. A comparison of the successive maps will show how great and numerous are the changes through which our group of islands have passed; and that, while we can recognise a closer approach in a general way to the physical features of the present day as time advances, there have been very wide variations of these features down to very recent times. The point which chiefly, perhaps, impresses the observer is, that the old mountain chains and groups of the north and west of Britain and Ireland when once formed, have ever after held positions of prominence, and were seldom, in some cases never, completely submerged. In a word, the north-western highlands of Britain, like those of Scandinavia, have been axes or fulcra for movements which have more powerfully affected the regions lying to the south of their position.

* The minute divisions of the Quaternary series are being carefully laid down on the Government maps of Belgium, "Commission de la Carte Géologique," under the department of the Minister of the Interior, but the scale is a large one, viz.: $\frac{1}{20000}$.

INDEX.

LONDON: PRINTED BY EDWARD STANFORD, 55, CHARING CROSS, S.W.